MARK DENNY

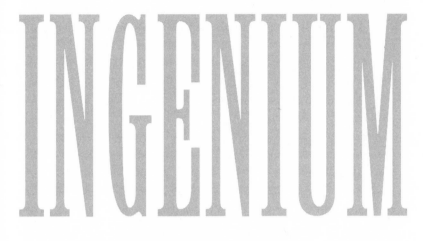

INGENIUM

FIVE MACHINES
THAT CHANGED
THE WORLD

THE JOHNS HOPKINS UNIVERSITY PRESS • *Baltimore*

© 2007 The Johns Hopkins University Press
All rights reserved. Published 2007
Printed in the United States of America on acid-free paper
9 8 7 6 5 4 3 2 1

The Johns Hopkins University Press
2715 North Charles Street
Baltimore, Maryland 21218-4363
www.press.jhu.edu

Library of Congress Cataloging-in-Publication Data

Denny, Mark 1953–
    Ingenium : five machines that changed the world / Mark Denny.
        p. cm.
    Includes bibliographical references and index.
    ISBN 13: 978-0-8018-8586-0 (hardcover : alk. paper)
    ISBN 10: 0-8018-8586-8 (hardcover : alk. paper)
    1. Inventions—History. 2. Machinery—History. 3. Force and
energy—Experiments. I. Title. II. Title: Five machines that
changed the world.
    T15.D3417 2007
    609—dc22   2006026085

A catalog record for this book is available from the British Library.

To Jane, my wife and best friend

CONTENTS

ACKNOWLEDGMENTS

The international nature of the historical machines in this book comes through clearly in this list of people from Europe and North America. I am grateful to each of them for assisting me, in one way or another, in writing the book.

Richard Adamek, steam engine enthusiast
Renaud Beffeyte, modern-day French medieval siege engine maker
Hector Cole, English arrowsmith
Dan J. Connelly, waterwheel enthusiast who likes to educate people about the historic Spencerville Mill, in Ontario, Canada
Erin Cosyn, acquisitions assistant at the Johns Hopkins University Press
Frans Dekkers, Dutch photographer—thanks for figure 3.3
Patrik Djurfeldt, Swedish maker of reconstructed medieval trebuchet
Sian Echard, Canadian professor of English

Rick Fairhurst, English steam engine photographer

Ed van Gerven, of the Dutch Windmill Society (De Hollandsche Molen)

Csaba Grózer, modern-day Hungarian bowyer

Cory Haugen, thanks for figure 2.1A

Ted Hazen (Pond Lily Mill Restorations), enthusiastic professional millwright from Pennsylvania

Kaare Johannesen, curator of The Medieval Centre in Denmark

JSTOR library services of University of Michigan, thanks for tracking down the original reference to Smeaton's waterwheel paper

Kinderdijk (Dutch village), thanks to your webmaster for figure 2.10.

Rob Langham, thanks for figure 2.1B

John Lienhard, American professor of engineering who maintains a website celebrating old machines

Vernon Maldoom, thanks for figure 2.11B

Amaury Mouchet, French physicist who appreciates ancient machines. Improved my longbow paper and recommended that I write one about siege engines.

Christoph Ozdoba, Swiss watch expert and enthusiast

Olivier Picard, Belgian medieval archery enthusiast

Amaro Rocha, of Digital Railroad

Sal, twenty-first-century saint, thanks for everything

Jeff Schierenbeck, of Wooden-Gear-Clocks, Wisconsin

Hugh Soar, of the Society of Archer-Antiquaries

Missy Staples, of Astragal Press, New York

Kurt Suleski, siege engine enthusiast

Paul Tarlowe, webkeeper for the Friends of Long Pond Ironworks, New Jersey

Ron L. Toms, manufacturer of model siege engines

Marc Tovar, of The Wooden Clockworks, Utah

Bob Vitale, manufacturer of modern-day waterwheels

Renato Zamberlan, of Antica Orologeria Zamberlan, Italy

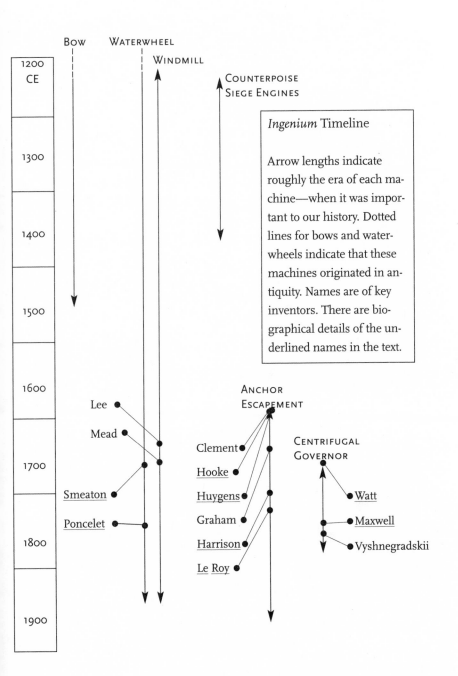

Bow Waterwheel

Windmill

Counterpoise
Siege Engines

1200
CE

1300

1400

1500

1600

Lee

Mead

1700

Smeaton

Poncelet

1800

1900

Anchor
Escapement

Clement
Hooke
Huygens
Graham
Harrison
Le Roy

Centrifugal
Governor

Watt
Maxwell
Vyshnegradskii

Ingenium Timeline

Arrow lengths indicate roughly the era of each machine—when it was important to our history. Dotted lines for bows and waterwheels indicate that these machines originated in antiquity. Names are of key inventors. There are biographical details of the underlined names in the text.

INTRODUCTION

The five machines I explore in this book are technologically simple. They are all mechanical, and the basic idea behind each of them is easily conveyed with a single picture. They are scientifically interesting and historically important. They still impact us today and they also leave less obvious but still discernible historical echoes. To give you an idea of what I mean by a historical echo, let me give you a trivial example.

There is a vulgar hand gesture with a meaning that seems to be more or less universal, but which takes on different forms in different parts of the world. In the United States it consists of a raised middle finger. In many parts of Europe it requires a whole fist and forearm to adequately convey the not-so-subtle nuances. In England the same gesture consists of two raised fingers, like Winston Churchill's victory sign but with the hand reversed, palm inward. This English exposition is a 650-year-old echo of the Hundred Years' War between England and France.

English archers were successful against the French cavalry during the Hundred Years' War, as we see in chapter 1. But woe betides an archer who was captured: he would have the index and middle fingers of his right hand cut off. Those two fingers were used to draw the bowstring—without them the archer was useless. When this early example of digital processing[1] became known, unmutilated English archers would taunt the French before a battle by showing them their two fingers ("I am a threat to you"). Such, at least, is the folklore. If true, the modern English gesture is an echo of history, albeit a distorted one.

Though trivial, this example demonstrates the effect of historical events on aspects of life far removed from them in both time and space. The machines examined in this book are important not only for their technical achievements but for their influence on the development of history.

So, this book blends history and technology in an examination of the development and influence of five machines. I show the physics behind the *bow and arrow*, the *waterwheel* (taken together with its close relative the *windmill*), the medieval *counterpoise siege engine*, the *pendulum clock anchor escapement*, and the flyball or *centrifugal governor*. For each device, my technical analysis has been published in a pedagogical journal. The journal papers arose because, without exception, all these devices contain useful and ingenious applications of physical principles of interest to those who teach physics. While researching the scientific development of the governor (the first of the five that I studied, though chronologically the most recent), I came across a lot of fascinating historical detail. Much of this detail, though, was inappropriate for the necessarily terse style of scientific papers, and so had to be omitted. The same was true for each of the other inventions. This stuff was *much* too good to gather dust—hence, this book.

But why *these* five machines? Why not, for example, the wheel or the lever? (Waterwheels and pendulum clocks require both these devices as building blocks, after all.) Well, it is certainly true that the wheel and the lever are historically important machines and that both of them provide elementary lessons for physics students. My studies, however, have been more concerned with the teaching of physics at a more advanced level—the journal which pub-

---

1. Sorry, this is a terrible joke—but you will have to get used to that.

lished my papers serves to provide university physics professors with clear and insightful new examples that they can apply when teaching their students. Here, in addition to "clear and insightful" we have a *Wow!* factor because of the indisputable importance of my chosen machines. Hence, the unusual combination in this book: technological history and classical mechanics. Yes, my machines teach us some pretty nifty physics, but they also changed the world.

The final chapter is motivated by the evident brilliance of the people, known and unknown, who were involved in the development and application of these five machines. I want to summarize the machines' impact, in their heyday and today, and to assess the inventiveness of the people who developed them. Much of their work predates the European enlightenment, from the sixteenth century CE, during which the modern concept of mathematical analysis of natural phenomena arose. (Until quite recently *physics* was referred to as *natural philosophy* in Scotland, a seat of the enlightenment in the eighteenth century.) Thus the development of these machines was largely empirical. No new laws of nature were unearthed. The craftsmen were not seeking a Nobel Prize, but rather a deer for dinner or a way to mine coal without drowning.

Equations. Ah, yes, I should say something about the level of mathematical presentation. Mathematics is the language of science, but not of most people, and so the author of a popular science book is faced with a dilemma. I must walk a tightrope here. It would be all too easy to lose the nonmathematical reader in a welter of baffling equations, or to bore the professional scientist with glib text that skirts around the technical details. So, I have gone to considerable effort to avoid falling off the rope. Where a key equation is presented, those of you who are not interested in the math can skip it without loss of continuity; the text gives an accurate explanation of physical principles. If, on the other hand, you like math then the key equations are here in place. I direct you to my technical papers for their detailed derivations—these papers contain a sound and comprehensive mathematical analysis.

Professional historians apart, there are three types of people: a few who don't *care* about history, many who don't *know* about it, and more than a few history buffs like me. The first group thinks that Java Man is a coffee seller, that the Ottoman Empire is a furniture outlet, and that the Enlightenment has something to do with the work of Thomas Edison. If you have read this far, then you are not in the first group, but probably fall into the second. Fear not, help is at hand. I don't intend to deliver a history lecture—I am not qualified to do that—but I do want to provide the context in which these five machines prospered. I find that a historical appreciation enhances my admiration for the inventors and their achievements, just as a picture looks better in the right frame, or a jewel in the right setting.

When compared with the Hundred Years' War example, echoes from the machines in this book are louder and clearer, or muffled and less distinct, but in all cases incomparably more important to our everyday lives. The bow has undoubtedly influenced human culture significantly, judging by its ubiquity. The Inuit peoples of North America used bows, though trees are scarce in their part of the world. Humans without bows might not have permanently occupied the Arctic regions. More certainly, the organized military application of bows led to the demise of heavy cavalry as the dominant force of medieval European warfare. Equally clearly, the waterwheel gave rise to modern turbines, thanks to the innovations of a nineteenth-century French engineer seeking to improve the efficiency of our oldest power generators—and win some prize money. Siege engines evolved into the gigantic and fearsome trebuchet, which changed warfare and hence history. Castles were built bigger and stronger to resist them, so improving the building skills and construction logistics of medieval Europe and the Middle East. More significantly, perhaps, trebuchets spurred the development of gunpowder weapons in Europe, with worldwide consequences. European maritime exploration and cartography were greatly enhanced when accurate clocks enabled longitude to be estimated with small error; accurate clocks were made only after the introduction of the anchor escapement. Steam engines powered the industrial revolution but were becoming unstable and inefficient, despite improved engineering, until the centrifugal governor was analyzed mathematically by one of the greatest theoretical physicists of the nineteenth century. Apart from considerable economic consequences, this analysis gave rise to the modern field of control engineering.

The influence of these five machines is felt today, because clever people applied science (whether they knew it or not) to harness nature. If you are a scientist, or someone who likes tinkering with machines, or if you enjoy poking your nose into the side alleys of technological history, then I hope that the following essays will stimulate. If so please tell me—especially if you know of other machines that are worthy of similar investigation.

Finally, I draw your attention to the footnotes. They add humor and educational details to the story that I tell in the main text.

INGENIUM

ONE

BOW AND ARROW

Early Days

THE BOW IS ONE of our oldest inventions. It was developed, we presume, to enhance our ability to hunt prey that was fleeter of foot than our ancestors, or perhaps was too wily or dangerous to approach within spear range. Whatever the reason, the bow proved so successful that it became nearly universal.[1] It spread across the ancient world, or was independently invented many times. From New Guinea to the Canadian Arctic, from South America to India, early man was able to establish himself in hostile environments with the help of this simple machine.

Simple? Well, yes, from a modern perspective. In its most elementary form, a bow consists of a single piece of wood. As we will see,

---

1. Aboriginal Australians may have adopted and then abandoned the bow and arrow. They had their own airborne weapon (Diamond).

though, considerable thought and effort went into the development, over many centuries and in many different cultures, of both bows and arrows. Despite its simplicity, this weapon was devastatingly effective. An arrow is accelerated at about 300g (g = acceleration due to gravity), and in a few milliseconds most of the energy stored in the drawn bow is transferred along the bowstring to the arrow. The animals that man had hunted for several hundred thousand years may have evolved to be wary of two-legged predators with spears, but they had no time to evolve strategies to avoid arrows. Death could reach them in a second from 50 meters away—a safe distance in the case of every other predator. A deer at 50 meters (50 m, or about 55 yards), or a larger animal at 100 m, was no longer safe.

Nor was an enemy. It is but a short step from hunting to warfare, and records show that bows have been used widely in various forms and in various tactical applications by military men through the ages. Jericho was probably defended by archers 9,000 years ago (Montgomery). Ancient Egyptians possessed sophisticated bows 4,000 years ago. Extensive historical records show that, 2,000 years ago, light infantry (skirmishers) preceded Greek and Roman heavy infantry formations into battle to unsettle and break up densely packed enemy formations—easy targets. Light cavalry performed the same function at speed, from the Parthian[2] horse archer in classical Persia to the Plains Indian of frontier America. The influence of the bow on history is most readily seen through the records of ancient and medieval warfare. If you grant me this assertion (and I accept that the bow also changed our history through hunting, but this process began in prehistory and is not recorded) then I can give a fairly precise date and time when this weapon changed the world: 13 August 1346 CE, sometime between 4 p.m. and dusk.

Crécy

EDWARD III OF ENGLAND must have been a worried man during the early afternoon of that fateful day, as he viewed the open landscape before him, from the vantage point of a windmill. He had invaded France the previous

---

2. The large Parthian empire blocked the eastward expansion of the Romans, with whom they were repeatedly at war. The Parthians were famous for their horse archers who were able, it is said, to shoot behind them at a pursuing enemy, while riding at speed. Hence, our phrase "parting shot" (Parthian shot).

month, and marched with his army of about 10,000 men toward Paris, before heading north to this small town in the Somme département of Normandy.[3] His force had been tracked by a much larger French army (perhaps 20,000– 30,000 men, depending on which source we read [Berners; Burne; Fowler; Harvie et al.; Seward]) under Philip VI. With his men tired and short of supplies, Edward decided to make a defensive stand and deployed his forces accordingly, on the low brow at the summit of a gentle slope, with woodland behind them and the enemy in plain view in front. The archers stood behind a forest of sharpened stakes, placed in the ground and pointing outward, to deter cavalry. But what cavalry! Edward saw from his windmill the cream of French chivalry, perhaps 10,000 mounted knights in plate armor, itching for a fight with the hated invader. They were supported by 10,000–20,000 peasant men-at-arms and 6,000 crossbowmen (Genoese mercenaries). Edward's army consisted of 7,000 English and Welsh longbowmen, and perhaps 3,000 knights, who dismounted to fight on foot, as befits a defensive stand.

The crossbowmen were no match for the longbowmen and soon retired in disarray (fig. 1.1). The crossbow could be fired only once or twice per minute, and the Genoese were in the open, without defensive pavises (shields, placed in the ground before them). The disorganized French knights were enraged by this reverse and charged impetuously toward the English lines, some actually riding down the crossbowmen in their anxiety to close with the enemy. They never made it that far, despite 14 or 15 repeated attempts.

Arrows "fell like snow" upon them "blotting out the sun," so thick was the cloud of arrows in the air. In the hands of a well-trained archer the longbow can be fired 10 or 12 times per minute. And Edward's archers were well trained. The yeomen of England were obliged by law to practice archery, and were said to be able to keep half a dozen arrows in the air at one time (this is rather unlikely[4]). They were strong men, too—skeletal remains show asymmetric growth, suggesting heavier musculature in the draw arm. The draw

---

3. This is a much-fought-over part of the world. In World War I, 570 years later, a much larger British army fared rather worse than Edward's did, but improved 28 years after that, in World War II.

4. The flight duration of an arrow fired to maximum range is about six or seven seconds. If the archer arrayed his arrows in front of him, for easy access and quick loading, he might be able to loose an arrow every three or four seconds. So, it is reasonable to suppose that a well-trained archer might be able to keep two, or at most three, arrows in the air simultaneously.

FIG. 1.1. *An 1839 rendition of the battle of Crécy, based on a medieval account of the battle. The English longbowmen are fighting Genoese crossbowmen (at a distance much compressed by the artist). I am grateful to Sian Echard for providing this image.*

weight of their bows has been estimated at an astonishing 80–110 pounds (Soar), giving a *cast,* or range, of about 250 m. It is widely considered that at Crécy half a million longbow arrows were fired (Bradbury; Hardy), and these rained down on the French cavalry at the rate of 2 tons per minute. The bodkin arrowheads penetrated plate armor when they struck it head on (nearly perpendicular), and the barbed broadhead arrows (see fig. 1.2 for a variety of arrowhead types) embedded in the flesh of the unarmored horses and men-at-arms. French casualties were at least 5,000 when darkness brought the battle to a close. (Exact figures are unknown, and estimates vary widely; 5,000 is a conservative value.) English casualties were limited to a few hundred. The

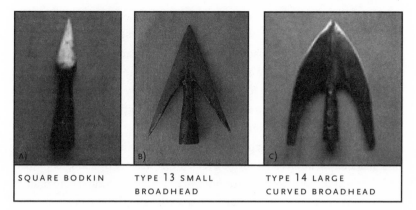

SQUARE BODKIN	TYPE 13 SMALL BROADHEAD	TYPE 14 LARGE CURVED BROADHEAD

FIG. 1.2. *Square bodkin arrowhead (left) and two broadhead arrows (type 13 small broadhead and type 14 large curved broadhead), made by Hector Cole, a modern-day arrowsmith. Used with permission.*

English knights had been little more than onlookers—this had been a stand-off victory, won at a distance by the archers.[5]

Evolution

CRÉCY WAS A NEW FORM of warfare. For several centuries previously, European battles had been decided by the feudal elite, armored knights who paid for their own mounts, armor, and retainers, and answered the call to battle by their king. Henceforth they fought on foot, at least until horse armor was developed. The days of chivalry died on the field at Crécy,[6] for these knights had been decimated by lowly yeomen, peasants who were paid by the day. For centuries such mounted knights had been accustomed to ruling the roost and had dominated set-piece battles and, indeed, feudal society. At Crécy the

---

5. Sigh. I must confess at this point, as an Englishman, to a slight feeling of guilt. Some time after my technical paper on this subject was published in the *European Journal of Physics*, I discovered that the reviewer employed by the journal for my article was a Frenchman. C'est la vie. However, instead of rejecting the paper in a fit of Gallic pique, he chivalrously suggested several useful changes that improved it.

6. Or perhaps a little earlier in this campaign, at Caen. In a well-recorded incident, a number of French soldiers expressed their displeasure at the invading English by exposing their backsides, in unison. Tragically, some of the soldiers underestimated the range and accuracy of the English archers. Ouch.

knights were beaten by massed ranks of trained archers, and they would be beaten again in this conflict, the Hundred Years' War, at the battles of Poitiers and Agincourt (the latter made famous by Shakespeare and, more recently, by Laurence Olivier and Kenneth Branagh). They were defeated by the rapid firepower of the longbow, widely regarded by military historians as the machine gun of its day, made most effective by being applied *en masse* by well-organized and well-trained archers.[7]

The surprise is that, perhaps contrary to popular opinion, the English longbow was not the pinnacle of the bowyer's art. It was not by a long shot (pardon the pun) the best bow that the ancient or medieval world had to offer. It was a good example of the rather ordinary *self-bow,* that is to say, one constructed from a single piece of wood, in this case, yew. The English longbow was without doubt cleverly designed for mass production, relatively robust, and easy to make, like a Kalashnikov rifle. Wood was cut from yew trees during the winter, before the sap rose. The *back* of the bow—furthest away from the archer when he fired—consisted of elastic sapwood, which is strong under tension, while the *belly*—nearest the archer—consisted of heartwood, which is stronger under compression. The bow cross section was roughly D-shaped, with the back being flat and carefully cut along the grain to avoid breaking the wood fibers. The construction took three or four years, with the wood worked in stages, before finally being strung. The English longbow string was made of hemp, extracted via a complex process from the fibrous stinging nettles.[8]

Clearly a great deal of thought and experimentation had gone into the construction of these bows, and in a prescientific age this development must have been empirical (trial and error). And yet even with all this developmental effort, bows developed earlier than the English longbow were more sophisticated and performed at least as well. In fact, the more complex *composite* bows appeared early in history.[9] These are made from several different materials,

---

7. With effective logistical support. It has been estimated that more arrows were manufactured for the English longbow than for all other bows.

8. Cannabis makes excellent bowstring, apparently, though it probably did not exist in medieval England. Presumably when it did later become available, it was not used primarily for this purpose . . .

9. If composite bows were better, then why were they not used by the English, who placed great importance on bows? Probably because the composite bow had not reached the northwest extremes of Europe. If so, then the English borrowed the long-

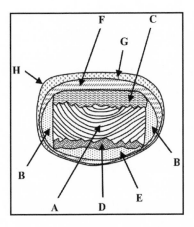

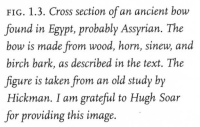

FIG. 1.3. *Cross section of an ancient bow found in Egypt, probably Assyrian. The bow is made from wood, horn, sinew, and birch bark, as described in the text. The figure is taken from an old study by Hickman. I am grateful to Hugh Soar for providing this image.*

each chosen for some specific quality or strength, and assembled to form the finished product. To give an idea of the complexity of these composite bows, consider the ancient example shown in cross section in figure 1.3. This Assyrian (or possibly Egyptian) bow dates from ca. 600 BCE. The areas labeled A and B represent different types of wood (Hickman 1959); C, D, and E correspond to horn, carefully shaped in short sections and glued into place; F and G correspond to sinew, teased out into individual strands from the leg or neck tendons of cattle, then glued into place; H represents an outer protective layer of birch bark. The horn, on the belly side of the bow, is elastic but strong in compression. As with the English bow nineteen centuries later, the wood is cut along the grain, for tensile strength. Again, we cannot escape the conclusion that a great deal of trial-and-error research and development (R&D) went into these machines, which tells us something about how important they were. Trial and error is a perfectly valid way to approach the problem—we should not disparage it simply because there is no theoretical analysis, which is so much a part of modern engineering R&D. Trial and error, properly conducted, is scientific and intelligent, in the sense of searching for and retaining measurable improvements in performance.

The Assyrians weren't the only people to develop composite bows; many

---

bow idea from the Welsh—and adopted it as their own—simply because it was the best bow they had encountered. Another reason might be the relative simplicity of construction of self-bows; recall that mass production and mass deployment were essential components of English longbow effectiveness.

cultures developed their own versions of this type of bow. Thus, Inuit peoples in the Canadian Arctic developed bows with sinew strung under tension along the back of the bow[10] (Baugh 1992). More southerly American Indians also made use of sinew to prestress their short bows, either by the Inuit method or by using glue (Baugh 1994). (The sinew is applied wet; it shrinks as it dries, which increases the effect.) The high point of ancient bow construction is often considered to be epitomized by the Persian and Turkish composite bows. These bows were made strongly *recurved*. That is, the unstrung bow curved away from the archer, sometimes to the extent that the bow tips met or crossed. Because of this recurving, the braced bow was under significant tension, thus increasing the draw weight and hence the power of the bow. I will show that recurved bows may be more efficient, and more powerful, than those that are not recurved. Turkish bows became specialized (as did their arrows, which must be "matched" to the bows); the *flight bows* were designed for maximum cast. Their longest recorded range is said to be over 800 m (Isles), though a reconstructed Turkish flight bow with a draw weight of 99 pounds tested in 1910 achieved a maximum range of "only" 434 m. The discrepancy is likely due to poor reconstruction; bow making was a lost art by 1910. The longer cast of 800 m is believable in light of the modern record range for a flight bow: an astounding 1,145.09 m (nearly three-quarters of a mile!). Admittedly, the modern record was achieved using a bow constructed from modern materials; nevertheless, it lends credence to the ancient claims. The current maximum range for the footbow and crossbow is over 1,800 m. A variety of bow types is shown in figure 1.4.

Modern Developments and Analysis

INEVITABLY, BOWS AND ARROWS eventually gave way to firearms. This displacement occurred in Europe in the sixteenth century CE. Consequently, in technologically developed countries interest in archery waned. A renewed curiosity in the early twentieth century led to a proliferation of archery clubs and associations. (There are over 50,000 members of archery organizations in the United States alone.) Competitions proliferate, though nowadays los-

---

10. This method requires less glue than is usually required for composite bow construction. The Inuit lacked glue, which needed significant amounts of a scarce resource—firewood—to make.

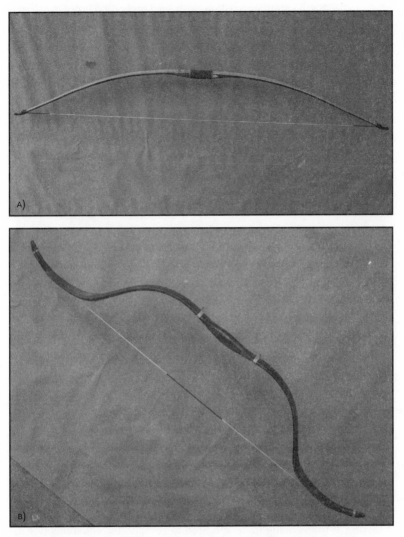

FIG. 1.4. *Modern bows of different types, made by a Hungarian bowyer, Csaba Grózer: a longbow (A) and an Indo-Persian recurve bow (B). Used with permission.*

ing an archery contest does not mean death in battle or going hungry. Modern archery equipment is hi-tech: carbon composite, spring steel, or fiberglass bows, carbon graphite or aluminum arrows, and nylon, Dacron, or kevlar bowstrings. The bows look different, too, when compared with their ancient equivalents. They are usually asymmetric, of the strictly modern *compound* design (a twentieth-century invention, with steel cables and eccentric wheels to make the draw easier). They have stabilizers jutting out, for smoother arrow release, and they are equipped with aiming sights. The draw weight averages 50 pounds (half that of the medieval English longbow). The compound design, aiming sights, and low draw weight all point to the modern archery contest, with the emphasis on accuracy, rather than range or hitting power. A modern archer must hold the bow steady while aiming, and this is easier if the draw weight is low. Also, the compound design means that most of the effort in drawing the bowstring occurs at the beginning of the draw—holding the string fully drawn is relatively easy. This is not the case for the ancient weapon in which we are interested. I guess that rate of fire and hitting power is more important than accuracy if your target is a hectare of perambulating cavalry.

One consequence of the upsurge of modern interest in *toxophily* (archery construction and practice) is that modern analytical methods have been applied to help bow makers understand the physics of archery. Prior to the 1920s, though, the scientific literature on archery is sparse. This reflects the fact that mathematical analysis techniques were developed only after the bow had ceased to be important. Much of the early scientific papers are devoted to experiment. Hickman developed a simplified mechanical model for a bow, which I have used as a basis for my own mathematical model (Denny). Despite the simplifications, Hickman's 1937 model is opaque. Schuster concludes that modern recurve bows approach 100% efficiency, a surprising claim that concurs with the earlier analyses. In the past quarter century, detailed numerical analyses by Marlow and Kooi reveal why simple models of bow physics lead to this claim, and also unveil the importance of bowstring mass and elasticity in reducing efficiency (the first of these was known to Klopsteg). These modern analyses, which treat the bow as a deformable beam of varying stiffness along its length, lead to complicated equations (for the technically inclined: six coupled partial differential equations, in the theory of Kooi) that cannot be solved exactly. They only yield answers when subjected to number crunching by a computer. The number-crunching mathematics tends to obscure the underlying physics. So, here you will find a simple model that captures realistic features of the more detailed analyses. I divide my explanation of longbow physics into two parts: internal dynamics (before the arrow is released), and external dynamics (arrow in flight).

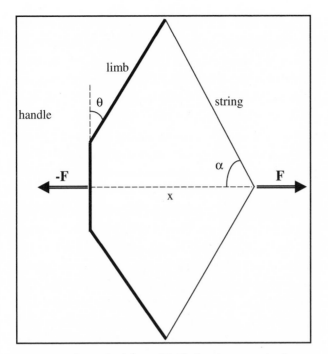

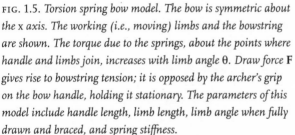

FIG. 1.5. *Torsion spring bow model. The bow is symmetric about the x axis. The working (i.e., moving) limbs and the bowstring are shown. The torque due to the springs, about the points where handle and limbs join, increases with limb angle θ. Draw force **F** gives rise to bowstring tension; it is opposed by the archer's grip on the bow handle, holding it stationary. The parameters of this model include handle length, limb length, limb angle when fully drawn and braced, and spring stiffness.*

Internal Bow Dynamics: Torsion Spring Model

WE CAN UNDERSTAND a lot about the physics of bow dynamics by mathematically modeling the bow in the simplified manner shown in figure 1.5. That is, we do not try to describe the true mechanical shape and bending characteristics of a real bow, but say that much of these characteristics can be captured by the simple torsion spring bow shown in the figure. This is a significant simplification that will enable us, figuratively speaking, to float over the forest canopy to our jungle destination (which I call "El Comprendo") instead of hacking through the mathematical undergrowth. A real bow, even a simple self-bow of the type used at Crécy, is a complicated problem in beam theory to a structural engineer. This is because the cross section of the bow varies with position along the bow; it is thicker near the *grip* or *riser* (handle) than at the ends. Thus the stiffness of the bow—how much it bends when the string is drawn, varies with position (less bending near the handle, more at the tips). It is not difficult to see that, were I to describe accurately the real bow in terms of beam theory, I would need a different model for each bow, depending on how the bow cross section changes with length. Instead, the torsion spring model (a user-friendly version of Hickman's model) enables us to see approximately what is going on, quite generally.

Assume (fig. 1.5) that a longbow behaves approximately like a perfectly stiff handle, attached to two perfectly stiff limbs. The limbs are each connected to the handle by torsion springs. When this bow is drawn, the angle between handle and limbs increases proportionally with the force of the draw. With this model, the equations that describe the internal dynamics simplify enormously. Putting in realistic numbers for draw strength, arrow mass, and so on, leads us to graphs like the ones in figure 1.6. (If you want the detailed math, you will find it in my 2003 article.) On the left we see how the arrow speed increases from zero at the instant that the archer releases the drawn bowstring, to about 60 m s^{-1}—say, 130 mph—a mere 17 milliseconds later. At this point the arrow leaves the string and "external dynamics" takes over. That is, the arrow speeds along its merry way, subject to the laws of gravity and aerodynamic drag (more of this later), before plunging into a modern bulls-eye target or perhaps a medieval French knight.

The second graph in figure 1.6 shows us something very interesting, because it is unexpected. Here the energy contained in the arrow, and the energy in the limbs of the bow, are plotted as a function of the bending angle (the graph looks almost the same if we plot energy against time). In this case, the fully drawn bow makes an angle of 45° between handle and limbs, whereas for the *braced* (undrawn) bow this angle is 20°. So, time progresses from right to left in this graph: the bowstring is released at 45°, the bow straightens up as much as it can, releasing energy stored in the torsion springs, until the limb angle is 20°. At this point all the energy put into the bow by the archer's arm has been released. But where has it gone? The graph shows that some of the energy goes into the limbs, but most of it goes to the arrow. As the bow straightens up, even the small amount of energy in the limbs is transferred to the arrow. So in this simple torsion spring model, 100% of the input energy is transferred to the arrow. The bow is *perfectly*

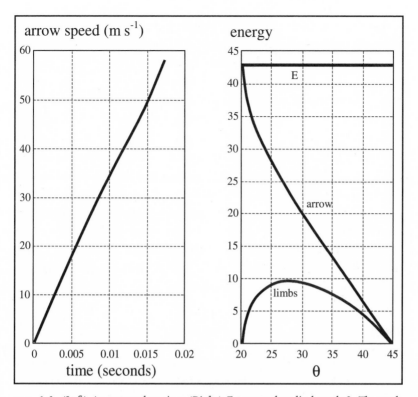

FIG . 1.6. (Left) *Arrow speed vs. time.* (Right) *Energy vs. bow limb angle* θ. *The total energy (the kinetic energy of the arrow and limbs, plus the stored potential energy of the torsion springs) is labeled E. It is constant here, because our model deliberately ignores dissipative effects such as friction. The limb angle in this case is 45° when the bow is drawn, as in figure 1.5, and reduces to 20° when the arrow is released (i.e., when the bowstring is braced, "undrawn"). Note that the limb energy reduces to zero when the arrow is released; the arrow takes away all the stored energy.*

efficient! This is a surprise. If we add in air resistance and other dissipative forces, we still find that the bow is very efficient (perhaps 98%). This high efficiency arises because frictional forces are not particularly strong (for example, the bowstring cuts through the air pretty well) and they do not have much time to act during the brief internal phase of bow-and-arrow dynamics.

So the simple torsion spring model is telling us that real bows can be very efficient machines. Given certain assumptions, the bow is 100% efficient for *any* bow model, not just the simple torsion spring model of figure 1.5. The main assumptions that lead to this unexpected result are:

1. Friction and air resistance are negligible.
2. The bowstring is inextensible and massless.
3. The arrow shaft is perfectly stiff.

Here "inextensible" means the string is inelastic—it will not stretch, no matter how much tension is applied—and "massless" means it weighs nothing. Clearly, these assumptions are idealizations—strictly untrue, but they are *approximately* true: frictional forces are small, the bowstring does not stretch much and it weighs very little, and the arrow is stiff (but somewhat flexible, as we will see). A lot of physics—and not just archery physics—consists of making simplifying assumptions like these, which permit general results to be obtained to a good approximation. Without the assumptions, we are back to hacking through the jungle. The trick is to make the right assumptions, so that the resulting model becomes mathematically tractable without jettisoning key features of the physics being modeled.

Internal Bow Dynamics: The Influence of Bowstring Mass

FROM WHAT I HAVE just said, you will rightly expect real bows to be very efficient machines. But of course they are not perfect. So, if our model tells us that bows are truly 100% efficient then one, at least, of the model assumptions must be wrong. Air resistance and friction within the bow (assumption 1) may reduce efficiency by 2% or so, and we expect that the small mass of the bowstring, and its extensibility (assumption 2), will also reduce efficiency. So, we now focus on the bowstring. Before going there, though, I draw your attention to the meaning of "100% efficiency." If all the energy put into the drawn bow by the archer is transferred to the arrow, then 0% of the energy is retained by the limbs (as you can see in fig. 1.6) or by the bowstring. That is to say, for a perfectly efficient bow, when the arrow is released, the limbs and the bowstring *are not moving.* In practice you might say: "Aha! I once fired an arrow from a bow, and I noticed what happens to the bowstring. Apart from slashing my left wrist because I forgot to wear the leather wristbrace, I noticed that the string quivers after the arrow flies off!" Indeed it does, and these bowstring vibrations take energy, thus reducing efficiency, but only if the string has mass. The effect of bowstring mass can be stated quantitatively as follows:

$$\text{Efficiency} = \frac{1}{1 + \frac{1}{3}\frac{m_s}{m}} \times 100\%$$

In this equation m_s is the bowstring mass and m is the arrow mass. For modern equipment a Dacron string has a mass of about 7 g (about one-quarter ounce) whereas the arrow mass is typically 25 g (under an ounce). This means that efficiency is reduced to about 91%, according to the equation. For medieval equipment, we would expect the string to be much heavier—a hemp string would have to be much thicker to possess the same tensile strength as Dacron—and we expect the arrow to possess a heavier (armor-piercing) arrowhead. Heavier arrows make the bow *more* efficient, but this factor is probably outweighed, as it were, by the heavier string. Recall that the draw weight of an English longbow was twice that of a typical modern bow, and so the string would have to be correspondingly stronger. Thus, medieval bows would be less efficient than modern bows, at about 70–80% efficiency, but this is still mightily impressive. (These numbers include all sources of bow inefficiency, and not just string mass.) So, medieval bows are indeed very efficient machines.

Internal Bow Dynamics: Bowstring Elasticity

IF THE BOWSTRING is stretchable, then bow efficiency is reduced. This happens because stretching an elastic bowstring causes energy to be stored in it, and later released when the bowstring contracts. This additional way of storing energy—in the string as well as in the bow—subtly changes the energy balance during the internal-dynamics phase of arrow flight. To understand why bowstring elasticity matters I need to explain, from the point of view of energy transfer, what happens when an arrow is fired.

For an inextensible bowstring, the stored energy is a minimum when the bow is braced. It is a maximum when the bow is fully drawn. When the archer's fingers release the bowstring, the potential (stored) energy in the bow is transferred to kinetic (moving) energy of the arrow and limbs. Like a ball rolling down a hill, the bow rapidly moves toward the minimum energy state that it can adopt, which is the braced position. At this point in the motion, the arrow separates from the bowstring. The arrow separates because the bowstring does not want to leave the minimum energy state, like a ball at the bottom of a hill.

Now, if the bowstring is elastic it turns out that the stored energy is no longer a minimum when the bow is in the braced position, but at a slightly different angle. For example, with the parameters of our torsion spring model in figure 1.5 we had a limb brace angle of 20°, and this angle represented the stored energy minimum. With an elastic bowstring the stored energy minimum occurs at about 20.5°. When the archer lets fly, the arrow now separates from the bowstring at 20.5° and not at the brace angle. So, compared with the ideal case, we find that the bow has 0.5° less "draw" and so does not quite transfer all the stored energy to the arrow. The residual energy (loosely speaking, the energy required to draw the string 0.5° in this case) is retained by the bowstring. Thus, a residual bowstring vibration remains after the arrow is released. (For a massless inextensible bowstring, recall that there was no bow motion at all once the arrow had separated.) So if the bowstring is elastic then not all the stored energy is transferred to the arrow. In other words, bow efficiency is reduced.

It is worth emphasizing that bowstring elasticity and bowstring mass are separate sources of vibration. An inelastic string with mass would vibrate, because inertia prevents it from coming to a sudden halt. If the bowstring is elastic then the string vibrates with larger amplitude, because it has a second reason to do so, as we have seen.

The "Archer's Paradox"

THIS INTERESTING AND well-known (to archers who happen to be engineers) problem straddles the divide that separates internal and external dynamics. It occurs after the arrow has parted from the bowstring, but before it has passed the bow handle. This phenomenon reduces bow efficiency, because it violates our third assumption. But I am getting ahead of the story. Here is the paradox.

The bow is usually held vertically.[11] However, because of the handle thickness, the arrow is not in the same plane as the bow and bowstring, because it must pass to the left or right of the handle. Here, to be specific, I will say that

---

11. Some Amerindian bowmen may be exceptions here, in that they held their bows in a horizontal plane when firing. This probably arose because they were hunting in an open environment, so that they needed to crouch, in low brush, to stay out of sight of their prey. Also some horse archers held their bows horizontally.

the arrow passes to the left. Now imagine yourself looking down on an archer as he fires. When he aims, he first fully draws back the bowstring, and then lines up the arrow with the target. So far, so good, but consider what happens when he lets fly. This is illustrated in figure 1.7. The string (and so the back end of the arrow, held to the string by a *nock*, or groove) moves toward the bow handle. It guides the arrow tail toward the center of the handle, and yet the arrow shaft must pass to the left of the handle. So, because of the handle thickness, the arrowhead is deflected to the left. Thus when the arrow begins its trajectory, it is off target. It should miss to the left. But the arrow doesn't miss, if it is lined up properly. How can this be? Your natural reaction may be to say that the archer compensates for this effect, by initially lining up the arrow slightly to the right of his intended target. But archers insist that they do not do this; they line up the arrow precisely on target. Yet it flies straight, and is not deflected by the handle. Interesting, yes?

The explanation (due to Klopsteg) is also interesting. See figure 1.7. The arrow is not perfectly stiff. When subjected to the sudden large force of the bowstring behind it, the arrow shaft buckles. It curves outward, so that the shaft does not touch the bow handle as it passes. The arrow vibrates in a horizontal plane (these vibrations quickly dampen during flight) so that, as the arrow tail passes the handle, it is buckled in the opposite sense and so the tail also does not touch the handle. Thus the arrow curves around the handle, as it passes, without touching it and so is not deflected from its original direction. This unexpected explanation resolves the paradox. (It also shows that not all the arrow energy is spent in propelling it forward through the air—some energy is wasted in these internal vibrations.)

The Archer's Paradox tells us that arrows must be matched to the bow that fires them; their vibration frequency must be approximately half the reciprocal bow release time. For the parameters of our torsion spring model in figure 1.5, this means that the arrow shaft must vibrate at about thirty cycles per second (30 Hz). If it does not, then some part of the arrow will come into contact with the bow handle as it passes, and so the arrow will be deflected. The vibration frequency of an arrow depends on several things: arrow length, shaft thickness, stiffness of the shaft material (ash or poplar wood, in the case of English longbows), and weight of the arrowhead. This matching of arrows to bows is well known to modern archers, and presumably was also appreciated by the archers of long ago.

The requirement that arrows must be matched to the bow that fires them

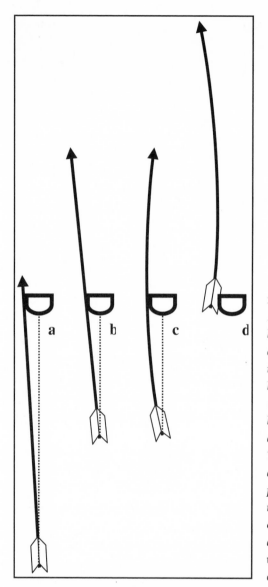

FIG. 1.7. The "Archer's Paradox." An arrow drawn back (a) and released (b) will change direction because of the width of the D-shaped bow handle, as the bowstring (dotted) is drawn toward the handle center. But in practice an aimed arrow will fly true. The resolution is shown in (c) and (d). The arrow is not perfectly stiff. It flexes under the strong compressive force of the bowstring, and curves around the bow handle without changing direction of flight.

is the clue that first alerted historians to the power of the English longbow. At the beginning of the twentieth century no historical examples of the medieval longbow were known, but a number of arrows had survived, simply because so many of them had been made. Knowing the arrow *spine* (stiffness) it was possible for some Sherlock Holmes historians to infer longbow power through the Archer's Paradox effect. In a nice example of history meshing well with science, these inferences were confirmed recently, when many late medieval longbows were recovered from the raised wreck of the *Mary Rose*. This large warship[12] had sunk in 1545 CE in the Solent (off the coast of southern England) during a battle with the French, under the baleful gaze of the English king, Henry VIII. The average length of the bows found on the *Mary Rose* was 2 m (6 feet 6 inches), and they were estimated to be able to penetrate mail armor at 200 yards. Their draw weight exceeded 100 pounds (Kooi 1997). These bows (and there were 137 of them in the *Mary Rose*, along with 3,500 arrows) were perhaps inferior to the longbows of the Hundred Years' War, a century earlier, since bows and arrows were giving way to gunpowder weapons by the time the *Mary Rose* sank. Nevertheless, they were still powerful killing machines.

External Bow Dynamics: What a Drag

SO NOW OUR ARROW has finally become free of the bow that launched it. It takes away about three-quarters of the energy supplied to the bow by the archer. Roughly half of the wasted energy is due to bowstring mass, and the rest is due to string vibration, arrow shaft vibration, and internal friction. The arrow leaves with a speed of about 60 m s^{-1}, as we have seen. How does this launch speed convert into arrow range?

I will begin by removing the earth's atmosphere,[13] at least from the vicinity of our flying arrow. It is then a matter of high school physics to derive, from Newton's Laws, the expression for maximum range of a projectile:

$$X_0 = \frac{v^2}{g}\sin 2\alpha$$

where X_0 is range, assuming that the ground is flat, so that the arrow lands at the same height it was launched. Launch speed is denoted v, launch direc-

---

12. There is an excellent Web site, www.maryrose.org/index.html, covering the details of this impressive ship, and the wealth of archeological information recovered along with her when she was raised in the early 1980s.

13. Yes, we physicists have that power. Now you know the type of person you are dealing with.

tion is at angle α above the horizontal, and g is the acceleration due to gravity at the surface of the earth ($g = 9.81$ m s^{-2} or 32 feet per second per second). Maximum range is obtained if the arrow sets off at an angle of α = 45° to the ground. If we take $v = 60$ m s^{-1} then we obtain a range of 367 m, or 400 yards. This range is further than the reported ranges for most medieval or modern bows (approximately 250 m). The difference, you will not be surprised to hear, is due to aerodynamic drag.

Drag is a complicated enough subject for simple, symmetrical objects such as cannonballs. For arrows it is more complicated still, because they are not aerodynamically simple; there are so many extra factors that come into play. For a cannonball, drag is well understood. The science of ballistics has reduced the complex physics involved to a single parameter, the dimensionless *drag coefficient*. This coefficient depends on the shape and texture of the cannonball surface, and varies slowly with speed, though for a given trajectory it can be taken to be constant. The drag coefficient encapsulates all the physical effects that give rise to air resistance. These are viscosity, pressure, and vortex formation and detachment. For cannonballs and other simple shapes we can say that air resistance increases as the square of launch speed, v^2, and increases proportional to the area that the projectile presents to the air as it rushes through the air. However, and this is where the pain begins, for more complicated shapes such as airplanes or arrows the situation is not so straightforward.

Aerodynamically an arrow consists of three parts: the arrowhead, the shaft, and the fletching (flight feathers).[14] I will go through these in order.

The arrowhead, usually made from steel, is the most variable part of the arrow (as you can see in fig. 1.2), because it has a different shape to fulfill different functions. We have already encountered the bodkin, which is a simple bullet-shaped point, used for hunting and for penetrating armor. It has the lowest drag of the different arrowhead types. The broadhead arrows are a class of flattened arrowheads, some of which are barbed, so that the arrow cannot be easily withdrawn from flesh or clothing. Other broadhead arrows are leaf shaped. This type of arrowhead has a larger and more variable drag, but some of them can also provide aerodynamic lift. (We will not consider the

---

14. In New Guinea, archers made fletchings from leaves. Historically, most of the world preferred feathers of one variety or another: goose, swan, eagle, pheasant, turkey. Nowadays fletchings are made from plastic.

whistling arrowhead, used in battles to undermine the morale of prospective targets.)

The shaft provides drag, to a degree that depends on the shaft length and diameter. Shaft drag force varies dynamically throughout the flight as the arrow direction changes. This variation arises because the arrow does not always point in the flight direction. It may "fishtail" (rear-end wobble, to put it in the vernacular), and it will turn over as the trajectory progresses. For example, if the arrow is launched at an angle of $+45°$ to the ground, then it will land point first at an angle of approximately $-45°$. (The archer wants this to happen, since he wants the arrowhead, and not the shaft, to strike the target.) So the arrow must turn in flight. This turning is assisted by the arrow weight distribution, with the center of mass (the balance point) usually 7% to 15% in front of the middle. (In the technical literature on arrow ballistics this is called the FOC, forward-of-center, value.) Different FOC values are chosen for target arrows, where accuracy is important, and for flight arrows, where distance is the main objective.

Most of the drag force that acts on the arrow comes from the fletching, depending on the size, shape, and number of "fletches." Fletching can be straight, twisted to the right or left, or helical (twisted a lot). Helical fletching causes the arrow to spin. Spinning helps with accuracy, since it evens out any asymmetric imperfections in the arrow, but it increases drag. Flight arrows usually have straight fletching, to minimize drag force.

Now please don't worry. I have no intention of taking you through the math. We have been saved from this by a physicist who happens to be interested in medieval archery,[15] and who has provided us with the following equation for maximum arrow range (Rees), given "average air":

$$X \approx X_0 \frac{1}{\left(1 + \dfrac{kv^2}{mg}\right)^{3/4}}$$

This parameterization captures all the complex drag behavior within the parameter k and is accurate to within a few percent. X_0 is the maximum range

<hr />

15. Gareth Rees works for the Scott Polar Research Institute (SPRI) of Cambridge University. The SPRI list of publications includes one on medieval archery—an unusual topic for such an institution, I would have thought.

in a vacuum, which we know already. The equation shows that maximum range in air decreases from X_0 as the square of launch speed, for low speeds, but decreases much more slowly as launch speed is raised. The arrow mass is m, and so mg is arrow weight. The value of k is about 0.0001 kg m^{-1} (so it is dimensional, unlike the usual drag coefficient).

Now we are in a position to estimate maximum range, including drag. Assuming that a medieval war arrow had a weight of 60 g (two ounces, more than twice the weight of a modern arrow) and a launch speed of 60 m s^{-1}, the equation gives us a range of about 250 m. This tallies with historical records and ties together our physical calculations with historical data. The calculations based on estimated bow power gave us launch speeds; taking aerodynamics into account has enabled us to convert these into longbow ranges. The ranges obtained agree with ranges reported by contemporary observers of medieval battles and by historians.

External Bow Dynamics: Lift

THE DRAW WEIGHT OF THE late medieval Turkish flight bows was in the vicinity 65–85 pounds. This draw is much less than that of the medieval English longbow, and yet the reported range of the flight bows is three times as far. Clearly, the Turkish flight arrows must have been designed to provide lift, to keep the arrow airborne.

The equation we have just utilized to calculate range assumes that drag is present, but it ignores lift. Neglecting lift is reasonable for most bow-and-arrow designs, but not for flight arrows where lift is crucial. Here I am again (and not for the last time in this book) faced with a dilemma. I want to explain to you the physics behind a complex phenomenon—in this instance, aerodynamic lift—without the heavy math. If you already know something about aerodynamics, then you know how mathematical it very quickly becomes, but then you don't need me to explain it to you. If you don't know any aerodynamics or fluid mechanics then I cannot explain it mathematically without writing a long book, and most of you would rather undergo root canal work than read it. So here I will aim for a happy medium and provide a simple, but not too simple, elucidation. We will float over the jungle.

Lift is an upward force that arises when an object moves through a fluid. The strength of the lift force depends on the type of fluid, here atmospheric air, and on the shape and speed of the object. A golf ball obtains lift due to the backspin applied to it, but this does not work here; instead, the arrow flies like an airplane. Consider, for example, how much effort has gone into designing airfoils during the past century, and you will appreciate (1) the difficulty in understanding aerodynamic lift and (2) the large lift force that can be generated if we get it right. It is not hard to conceptualize that an arrow may be designed to generate lift. A broadhead tip and long fletchings may act as an airfoil. Recall that the arrow does not necessarily point in the direction of travel. If it is moving horizontally, and yet is tilted slightly upward, then you get the idea. Compare this arrow with a jet fighter that has canards (little wings) at the front and larger wings further back. Replace the canards with the broadhead tip and wings with fletching, and we have the potential for an arrow with lift. Recall also that the aspect presented by the arrow (the angle between flight direction and pointing direction) is to some extent determined by the FOC number, which for flight arrows can be chosen to maximize lift.

I offer the following equation to illustrate the effects of lift on achievable arrow range. In footnote 16 I outline, in brief, how it was derived. If you are not interested in derivations then just trust me. Anyway, the derivation is just mathematics, not physics, and I will now explain the physics that went into it. First, though, here is the equation for arrow range assuming drag and lift:

$$X \approx X_0 \left(1 + \frac{f_L - 4f_D \sin\alpha}{mg} \right)$$

Here X_0 is the range in a vacuum, as before. $f_{L,D}$ are the lift and drag forces, respectively, mg is arrow weight, and α is the launch angle of the arrow. Recall that, for maximum range in a vacuum, this must be 45°. For our equation $\sin\alpha$ indicates that the optimum launch angle is a little less than 45°—in

---

16. The arrow range is assumed to be a function of speed, and is expanded in a Taylor series about the vacuum value, neglecting the second and higher derivatives. The increment dv is taken to be $dv = -a \cdot dt = -aT$, where a is drag force divided by arrow mass (since drag opposes velocity direction), and T is flight duration. Replace g by $g-b$, where b is lift force divided by arrow mass (since lift is, on average, directed upward). Assume lift and drag forces are small compared with arrow weight and keep only first-order terms. This is rough-and-ready engineering mathematics, with many sweeping and extravagant approximations—it would make a pure mathematician curl up and die. But planes fly, nevertheless, and some of them carry mathematicians.

agreement with observations of real projectiles. The main simplification I have used in the derivation is to assume that lift and drag forces are constant. The lift force points skyward and the drag force points backward. In reality they both depend on arrow speed and aspect, both of which vary throughout the flight, but it is a reasonable approximation so long as we do not expect the equation to be precise. It is approximate, and we can expect to draw from it only qualitative conclusions.

Given this, what does my equation tell us? First, see what happens if we ignore lift. This leaves drag, and we see that maximum range is reduced. (Further, by equating my equation—minus lift—with our earlier equation we find that we get the right kind of expression for the drag force. For example, it tells us that the drag force increases as the square of arrow speed.) Also, as with the previous equation, we find that light arrows worsen the effect of drag. Now add lift. We see that if the arrow can generate enough lift force to overcome drag, then maximum range increases. Flight arrows must be able to generate large lift forces, forces that are comparable in magnitude with arrow weight, since the quoted flight arrow range is more than twice the range X_0 that is attainable in a vacuum.

Recurve Bows

I MENTIONED RECURVE bows earlier. With prestressing, the bow acquires more draw weight than a standard bow made of the same material. Put another way, a recurve bow with the same power as a standard bow would be shorter, and therefore more manageable. This would be important for horse archers, or for hunters in dense forests where maneuverability is compromised. A strongly recurved bow is illustrated in figure 1.8. See also figure 1.4.

These recurve bows are more efficient than standard bows. This extra efficiency arises because the length (and so the mass) of bowstring that vibrates, upon releasing the arrow, is reduced. We have seen how vibrating bowstrings reduce efficiency. There are two reasons for the reduced length of vibrating string. First, recurves of the same power as standard bows are shorter. Second, you can see from figure 1.8 that the braced recurve bow has part of the string lying adjacent to the bow limbs. This part of the string will not vibrate when the arrow is released—only the "free" string can move in that way. Thus, recurves are more efficient. (Modern recurve bows may be close to 100% efficient [Schuster].)

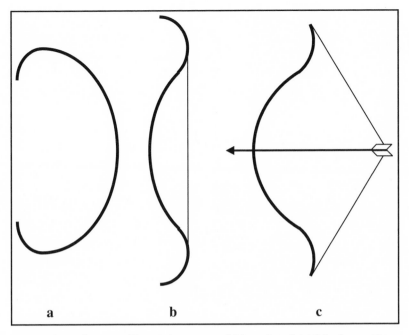

FIG. 1.8. *Strongly recurved bow: (a) unstrung, (b) braced, and (c) fully drawn. Note that, depending on the draw, a variable length of the bowstring touches the bow, and so is stationary when the limbs are stationary.*

The improved convenience and efficiency of recurve bows comes at a high price, literally. These bows take longer to construct, because the construction process is more complicated. Also, recurve bows were often (though not always) composite, which are more complicated to make anyway. Thus, composite recurve bows, such as the ancient Persian or Turkish bows, represent the high-water mark of a historical bowyer's art. Or science.

Catapults

THE WAR ENGINES of classical antiquity, both Greek and Roman, included torsion-powered catapults that may be described mathematically by our torsion spring model of the bow. The ancient Greek *euthytonos* was far heavier than a hand-held bow, and was mounted horizontally on a swivel, which was itself mounted on a substantial tripod frame. The two working limbs consisted of straight sections of thick wood, which were connected to the sta-

FIG. 1.9. *Modern reconstructions of two types of torsion catapult. The horizontal two-arm ballista (A) has separate torsion springs for each arm, clearly visible here, whereas the vertical arm onager (B) has a single arm powered by twisted rope. Thanks to Kurt Suleski, Knight's Armoury, for permission to reproduce the ballista picture, and to Ron. L. Toms for permission to reproduce the onager.*

tionary frame (equivalent to the bow handle) via springs made from twisted animal tendon. Sometimes horsehair or human hair, which is quite elastic, would substitute for sinew.[17] Experiments conducted on modern reconstructions of such catapults showed that they twisted in the same elastic manner as our torsion springs (Landels), so this model may be applied to ancient catapults. See figure 1.9 for examples of these classical torsion engines.

It is quite common in physics for the same mathematical theory to apply to seemingly different applications. For example, the so-called simple harmonic motion (SHM) is ubiquitous. It applies approximately to the oscillations of a pendulum swing, to the vibrations of atoms in a molecule, and to a host of other phenomena. All that is required is to change the SHM parameters and *voila:* one theory fits all. Here we find that the theory of bows and arrows applies to classical catapults.

In our case the parameters that need changing are the handle, limb and bowstring sizes and masses, and the torsion spring constant. You will not be surprised to hear that all are larger for the catapult than for the bow. Projectile mass is also larger. The catapults were highly prestressed, by virtue of the twisted sinew. One difference is that the limbs were held in the brace position not by tension in the bowstring, but rather by vertical stops attached to the frame (this is evident in fig. 1.9). These were padded so that, when a drawn bow was released, the limbs did not break when slamming into these stops. (Nevertheless, broken bow limbs were the most common cause of catapult failure.) Presumably the *raison d'etre* for the stops is that the bowstring was not strong enough to withstand the huge tensions that arose when the catapult was fired and reached the brace position—the string would snap. The catapult bow was drawn using ratchets applied independently to each spring (in the case of two-armed catapults). The catapult was then ready to loose its heavy arrow. Larger versions (*palintonos*) fired stone projectiles.

From our experience with the torsion spring bow, we can see that these catapults would have been relatively inefficient. This is because the limbs would not transfer all of their energy to the missile, and so be brought to a natural halt. Instead the limbs would be halted by the stops, releasing the missile before having transferred all the stored energy. Indeed, the throwing arm(s) will attain their greatest speed just as the projectile is being released, so that much

---

17. In the siege of Carthage, 148 BCE, women of the city sacrificed their hair to make catapult springs.

of the energy stored in the twisted rope or sinew is transferred to the arm, but not to the projectile. Additionally of course the violent impact of limb on stop would generate heat and sound energy, further reducing efficiency, as well as catapult life.

So, these machines lacked the efficiency, as well as elegance, of the archer's bow. Perhaps this is why the bow survived from classical antiquity through to the late Middle Ages, whereas these old catapults were eventually replaced by more efficient and much larger counterpoise engines, which form the subject of a later chapter.

Leftovers

THERE ARE A FEW ASPECTS of bow dynamics that I have not touched on. I will simply list them here, to give you an idea of the extra physics that they introduce. In general, these phenomena complicate the analysis, but do not significantly alter the conclusions that we have drawn from the simple torsion spring model.

- *Gravity.* If a bow is held vertically, and drawn with the arrow pointing more or less horizontally, then gravity will influence the two limbs of our torsion spring model differently. Gravity will assist the lower limb in returning to the brace position, and deter the upper limb (see fig. 1.5).
- *Aymmetry.* Some bows were designed with the two limbs—above and below the handle—of different size or shape. To model this we would need to assume different torsion spring constants for each limb. Aiming such a bow would be trickier than aiming a symmetric bow.
- *Handle stability.* I have assumed that the bow handle is held rock steady by the archer's left arm (assuming he draws with his right). In practice this may not be so, since the archer's body is flexible, and will react as the bowstring is released. His arm may buckle, and his body may sway slightly, or twist, under the influence of the large recoil.
- *Hysteresis.* One complication that arises in considering the detailed analysis, when the bow is considered to be a continuously deformable beam, is the difference between static and dynamic deformation. To understand what this means, think of a single frame taken from a high-speed film of a bow being fired. Say the arrow is halfway from draw position to release. The shape (deformation) of the bow may not be the same as it would

be if the bow was drawn to this position, and held there. Consult Kooi, if you are interested in learning more about dynamic deformation of bows.

REFERENCES

Baugh, R. A. (1992). *Bulletin of Primitive Technology* 1:3.

Baugh, R. A. (1994). *Bulletin of Primitive Technology* 1:7.

Berners, Lord, trans. (1904). *The Chronicles of Froissart*, ed. G. C. Macauly. London: Macmillan.

Bradbury, J. (1992). *The Medieval Archer.* Rochester, NY: Boydell and Brewer.

Burne, A. H. (1955). *The Crécy War.* London: Eyre & Spottiswoode.

Denny, M. (2003). *European Journal of Physics* 24:367–378.

Diamond, J. (1997). *Guns, Germs, and Steel.* New York: W W Norton.

Fowler, K. (1967). *The Age of Plantagenet and Valois.* New York: C. F. Putnam.

Hardy, R. (1993). *The Longbow.* New York: Lyons and Burford.

Harvie, C., C. Kightly, and K. Wrightson (1980). In *The Illustrated Dictionary of British History,* ed. Arthur Marwick, p. 80. New York: Thames & Hudson.

Hickman, C. N. (1937). *Journal of Applied Physics* 8:404–409. Hickman's model is also analyzed in Marlow, W. C. (1981). *American Journal of Physics* 49:320–333.

Hickman, C. N. (1959). *Journal of the Society of Archer-Antiquaries* 2:21–24.

Isles, F. J. (1961). *Journal of the Society of Archer-Antiquaries* 4, available at: http://www .student.utwente.nl/~sagi/artikel//turkish/

Klopsteg, P. E. (1943). *American Journal of Physics* 11:175–191.

Kooi, B. W. (1981). *Journal of Engineering Mathematics* 15:119–145.

Kooi, B. W. (1991). *Journal of the Society of Archer-Antiquaries* 34:1.

Kooi, B. W., and C. A. Bergman (1997). *Antiquity* 71:124–134.

Kooi, B. W., and J. A. Sparenberg (1980). *Journal of Engineering Mathematics* 14:27–45.

Kooi, B. W., and C. Tuijn (1992). *European Journal of Physics* 13:127–134.

Landels, J. G. (1978). *Engineering in the Ancient World,* chap. 5. London: Constable.

Montgomery, B. (1972). *A Concise History of Warfare.* London: Collins.

Rees, G. (1995). *Physics Review* 4:2–6.

Schuster, B. G. (1969). *American Journal of Physics* 37:364–373.

Seward, D. (1978). *The Hundred Years War.* New York: Atheneum.

Soar, H. (2004). *The Crooked Stick: A History of the Longbow.* Yardley, PA: Westholme.

TWO

WATERWHEELS

AND WINDMILLS

I PLACE WATERWHEELS and windmills in the same chapter, count-ing them as one machine for the purposes of this book. Of course, the engineering requirements of waterwheels and windmills are different, but the physical principles are similar. Both machines con-vert natural fluid flow into usable power, initially in the form of torque. First, I'll share with you the delightful waterwheel, because it is the older and more widespread of our two ancient *prime movers* (power sources). Windmills arrived on the scene considerably later than waterwheels, and the most developed of them were capable of generating greater power. Undoubtedly, though, waterwheels were more important on a global scale, because they were so common in so many countries for so long.

Our Powerhouse for Two Thousand Years

OUR OLDEST MAN-MADE power source,[1] waterwheels, was widespread across the Old World by the first century BCE. Waterwheels have not been a part of human history for as long as bows and arrows, but they have played an important role in *recorded* history for longer (see the Timeline). The fact that latter waterwheel development has been recorded is important for the historian of technology, who seeks to understand the evolution of waterwheels from simple beginnings to modern hydraulic turbines. So, our oldest power-generating machine transforms into one of our most modern.

It is hard to overstate the historical importance of waterwheels. Their development over the millennium from 500 CE to 1500 CE represents mankind's most outstanding technological development of this period (Britannica). Waterwheels powered most of the Old World civilizations during this millennium and it is difficult to imagine what the world today would be like, had we lacked these machines during this formative period. In the next section I describe the different types of waterwheel. We will see how the enduring undershot and overshot versions evolved from the older Norse wheel and how they turned, as it were, into water turbines. The emphasis will be on undershot and overshot wheels because these designs occurred at important crossroads in the history of technology. Indeed, they helped to *define* these crossroads.

Waterwheel numbers increased substantially during the Middle Ages, when there was an acute shortage of labor, making laborsaving machines more cost-effective. It is widely considered (Britannica; Mason; Usher) that the most dramatic industrial consequences of waterwheels occurred in the Middle Ages, when the scale of milling increased considerably with the development of large towns.[2] The considerable economic and social impact of waterwheels may be judged by their increased application (Gies; Dresner).

---

1. Draft animals, oxen and horses, were an older and more common, if temperamental, power source that survived into modern times.

2. Indeed, we might turn this statement around and say that large towns developed where waterwheels operated. The shortage of labor in Europe during the Medieval period was due to the devastating Black Death (bubonic plague, which spread westward from China in the fourteenth century, killing one person in three). Fewer workers led to higher wages, which provided an incentive for labor-saving innovations such as improved waterwheels.

FIG. 2.1. *(A) Without the assistance of wind or water, mills were powered by humans or draught animals. This mill appears to be set up for human muscle power. (B) Undershot waterwheel, providing power for a watermill in England. Thanks to Cory Haugen for the millstone picture (A), and to Rob Langham for the waterwheel shot (B).*

From grinding grain (see fig. 2.1) and pumping water in antiquity, water-powered mills were later developed to forge iron, full cloth, saw wood and stone, and perform metalworking and leather tanning.[3] Later, waterwheels were ap-

---

3. One serendipitous use of waterwheels in the Middle Ages was as sewage treatment plants. A waterwheel that was located downstream of a population center, such as a monastery, could contain a lot of bio-waste, unfortunately for the miller. By aerating the water, the waterwheel acted to treat this sewage (Magnusson).

JOHN SMEATON (1724 – 1792)

Born in Austhorpe near the city of Leeds, in Yorkshire, England, John Smeaton was the son of a respected lawyer. As a child he attended Leeds Grammar School. His father expected John to follow in his footsteps, and to this end he sent John, aged 16, to London to study law. However, John was much more interested in the scientific meetings of the Royal Society than he was in studying law. In 1748 he opened a shop selling scientific instruments, with the reluctant acquiescence of his father. He published technical papers on electricity (this novel phenomenon was all the rage among the scientific "philosophers" at this time) and astronomy before becoming interested in engineering in 1752. Seven years later, in May 1759, he presented to the Royal Society his great medal-winning paper, published under the title "An Experimental Enquiry concerning the Natural Powers of Wind and Water to Turn Mills, and Other Machines, Depending on a Circular Motion." The title reveals his method: careful experimental investigation. Matthew Boulton, whose later business partnership with James Watt proved so influential, consulted Smeaton about watermills for his factories. Also in the 1750s Smeaton took on the project of reconstructing the Eddystone Lighthouse. All previous constructions on this storm-lashed site off the southwest coast of England had failed. Smeaton built his lighthouse of interlocking stone blocks, held together by a new type of mortar (which he developed, again following careful experimental investigations) that would set under water. His Eddystone Lighthouse lasted for 100 years.

Following these successes Smeaton was called upon to organize and manage the construction of many building projects—canals, bridges, harbors, and piers—that represent the beginnings of England's gearing up for the industrial revolution. He designed fifty watermills, windmills, and steam engines. His interest in steam power began in the 1770s; he doubled the efficiency of the early Newcomen atmospheric engine and built two colliery steam engines that, between them, were the most efficient and the most powerful of any engines, prior to those of his contemporary, James Watt. Smeaton coined the term "civil engineers" (to distinguish them from military engineers) and founded the Society of Civil Engineers in 1771. He died in his garden the next year, following a stroke.

plied to drive the machines of the early industrial revolution. Thus—and this is only one example of many—Josiah Wedgwood, the pottery magnate and a driving force behind the early industrial revolution, utilized waterwheels to power his factories in the 1770s (Uglow).

The power of European waterwheels increased by a factor of three during the eighteenth century, to perhaps 10 kW. This improvement became possible because millers and millwrights of this century began to adopt a more systematic and scientific approach, characteristic of the age, to mill design and construction. Because of their importance, much effort went into the scientific investigation of waterwheel efficiency during this period. In 1704 Antoine Parent calculated the maximum efficiency of an idealized undershot waterwheel. In England, between 1752 and 1754, John Smeaton (founder of the Society of Civil Engineers) made scale models of both undershot and overshot waterwheel designs (Smeaton 1759). He varied components one at a time, patiently and laboriously, to empirically establish the most effective designs, and he concluded that undershot wheels were no more than 22% efficient, whereas overshot wheels were 63% efficient. In 1780 the influential Swiss mathematician Leonhard Euler studied the latest waterwheel developments. In the early nineteenth century J. V. Poncelet, a French mathematician, increased the power of undershot waterwheels to that of overshot wheels (Britannica; Mason). The famous (to physicists) 1835 paper by Coriolis was written, not on the subject of earth rotation,[4] but rather on energy transfer in rotating systems such as waterwheels (Coriolis).

It amazes me that such a significant nineteenth-century scientific and engineering effort went into improving an instrument first introduced nearly two thousand years earlier. This fact provides a clear illustration of the absolute indispensability of waterwheels to the development of European and New World commerce and industry.

Wheels within Wheels

THE HISTORICAL AND archeological evidence for early waterwheels is surprisingly scant. We cannot say where the earliest type—the Norse wheel—

---

4. The "Coriolis force" arises in rotating systems, such as the earth, and is best known in meteorology because it gives rise to trade winds, determines the direction of rotation of cyclones, and greatly influences ocean currents.

originated. It is not named for its place of origin, but because many have been found in Scandinavia; it is possible that geographical conditions there favored this type, since these wheels require fast-running water to provide torque to the paddle vanes, and so they work better in mountainous terrain. We do know that the old Norse wheel existed in many geographically separated places, including China and Greece, by the first century BCE. The Roman writer Vitruvius gave us a good description of them in 27 BCE, so we know that these wheels were common in the Roman world at this time.[5]

The Norse wheel consists of a paddle wheel attached to the lower end of a vertical axle. The upper end is attached to whatever implement requires rotary power. The *headrace* is the source of water: a sluice or flume that brings the flow to the paddle. The *tailrace* removes water from the paddle. The difference in height between these two is known as the water *head*. In Europe, one of the main uses of waterwheels through the ages has been to grind corn into flour. In this case the axle passes up through the center of a stationary circular *bedstone*, and attaches to a similarly sized and shaped *runner stone*. The corn is placed between these two, and the heavy runner grinds it into flour. Norse wheels are the simplest waterwheels because they require no gearing. The gear ratio is 1:1, so that the runner stone rotates once for every turn of the paddle.

The advantage of this early waterwheel was simplicity of design and construction. Because there were no gears, there was little loss of power along the axle connecting paddle to grindstones. There, however, the advantages of the Norse wheel end. The disadvantages are clear. First, because the gear ratio is low, the Norse wheel needs fast-flowing water, and lots of it. To obtain this supply of fast-flowing water the wheel was sometimes placed at the bottom of a pit, with a sluice to remove the "spent" water. But such a pit limited the size of the paddle, and so limited the wheel power. In practice Norse wheels were small, providing power for one family or perhaps a few families at most. Second, the absence of gears meant that the speed of the grindstone was completely determined by the rate at which water flowed across the paddle vanes. Since streams and rivers do not have constant flow rates, the most efficient grindstone speed could not be maintained. Fluctuating grinding

---

5. Because of this historical source, the Norse wheel is sometimes called the Vitruvian wheel. Also, confusingly, it is has been named the "vertical shaft wheel," the "horizontal" wheel, and the "tub" wheel.

speed was bad news for millers because it meant that the mill output, flour say, was of variable consistency. I discuss this point further in a later chapter, because the need for constant grinding speed led directly to another of our world-changing machines.

To overcome the limitations imposed by the simple Norse wheel design, the wheel was turned 90° and gears were introduced to yield the undershot wheel. Here the paddle wheel is vertical, with the lower part placed in a running stream. The horizontal paddle axle is connected to a vertical grindstone axle via gears, typically a toothed wheel meshing with a lantern pinion (Landels). Gone were the pit and the mountain streams. Undershot wheels worked (and still work) well in sluggish water with low flow rates. The wheel can be made large, to increase the torque supplied by water flow. Some power is lost by the gearing, to be sure, but this is more than compensated for by the increased torque of the larger paddle, and by the flexibility of gear ratio $1:n$. Here n could be changed as water flow rate changed, so that optimum grindstone speed was maintained. Since undershot wheels worked in slow streams they could be erected in many geographical regions that lacked fast-flowing water, such as the flat plains upon which most classical civilizations developed. So undershot wheels proliferated, and remained the most common type of waterwheel from classical antiquity until overtaken by overshot wheels in the thirteenth century CE (Britannica). Even after the development of the overshot wheel, they remained popular right up until steam engines became efficient and widespread in the middle of the nineteenth century CE.[6]

Overshot waterwheels existed in the Roman world, and may have been a Roman development.[7] Here the paddle wheel is vertical, as with undershot wheels, but the water falls on top of the paddle. So, imagine looking at two wheels, undershot and overshot, with water flowing from left to right. Water passes beneath the undershot wheel, turning it counterclockwise, and passes over the top of the overshot wheel, turning it in a clockwise direction. The overshot design is intrinsically more efficient—it extracts three times as

---

6. "Dedham Lock and Mill" is a well-known oil painting by John Constable, who idolized rural life in early nineteenth-century England. It depicts a mill powered by a large undershot waterwheel.

7. Perhaps they were inspired by the bucket wheel, used to raise water from a river, and powered by humans treading on the wheel. Reverse this process to yield an overshot water wheel: falling water is used to create useful power in the form of torque.

FIG. 2.2. A nineteenth-century overshot waterwheel. This one happens to be in West Virginia (United States) but it could be from anywhere in the Old or New Worlds. Note, in particular, the headrace, a sluice constructed to bring water to the top of the wheel, and the large water height difference between the incoming and outgoing flow. This photograph illustrates that overshot wheels take advantage of the water weight, as well as water flow speed, to turn the wheel. I thank Ted Hazen of Pond Lily Restorations for permission to reproduce this image.

much energy from a given source of water—because it makes use of the water weight as well as flow speed. Both weight and speed contribute to paddle-wheel torque. An overshot wheel is shown in figure 2.2.

Given the threefold superiority of overshot efficiency, why persist with undershot waterwheels at all? I have already stressed the importance of waterwheels as a source of power since antiquity. The Domesday Book of 1086 CE recorded more than 5,000 mills in England—that corresponds to one wheel for every 400 people. By 1820 France alone had 60,000 waterwheels. The dense population of mills along early nineteenth-century European rivers and streams meant few hydro sites remained available, so water head became a scarce and valuable resource. Real estate, in this case running water frontage, became ever more expensive. Water head was squabbled over, and every last foot was needed. Overshot wheels required a large head—the drop in height must be about the paddle diameter, between 2 m and 10 m—and so were usually confined to hilly areas, or required extensive and expensive auxiliary construction, such as mill races that ran for hundreds of meters. Undershot wheels, on the other hand, could operate with less than 2 m head and so could be located on small streams in flat areas, near population centers. Thus they remained important well beyond the period when scientific investigation had shown them to be relatively inefficient.

I have already mentioned the second reason for the persistence of undershots. Recall that in the nineteenth century Jean-Victor Poncelet increased the power of undershot waterwheels to that of overshots. The French government offered large prizes for improved waterwheel design to boost economic development in industries thirsty for power, and this prize money spurred a lot of theoretical and experimental investigations. Poncelet won a prize with his study, published in 1826 CE, which recommended modified waterwheel vanes, and these vanes proved to be an immediate success. Poncelet's modifications anticipated the future development of hydraulic turbines. They improved undershot waterwheel efficiency by a factor of three, so that undershot wheels became as efficient as overshot wheels.

Waterwheel Efficiency: Idealized Overshot Waterwheel Model

SMEATON'S CAREFUL STUDY showed that overshot wheels were about 65% efficient. That is to say 65% of the water power input to these machines was converted into useful power, to turn a grindstone, or lathe, or potter's wheel, or to operate a trip hammer. He showed that undershot wheels were only about 22% efficient. In the next few sections of this chapter I am going to show you why these figures apply. It turns out that we can devise a simple mathematical model of waterwheel physics that reproduces these figures, determined empirically pretty well by Smeaton nearly 250 years ago. What's more, the model shows us how Poncelet's modifications led to a great increase in undershot wheel efficiency. This analysis sheds light on the engineering principles and design features that exercised the minds of many well-known and unknown scientists, engineers, and millwrights through the ages.

The physics that I will convey to you involves energy conversion, fluid mechanics, and, above all, torque. I start with a simple model, and then build in more and more realistic features. This building process is necessary here, but was not needed in chapter 1, because waterwheels are much more complicated machines than longbows are, with more moving parts, more variability of components, and more structural variations. So I must seek to present a simple-but-not-too-simple core model, which catches the mechanical essence of waterwheels, upon which we can later hang various bells and whistles to fine-tune our predictions.

Consider the idealized overshot waterwheel shown in figure 2.3. This is our simplest possible overshot case study. We suppose that the wheel has several buckets attached to the rim and these buckets are free to rotate without friction. Water drops into the buckets vertically, from a spout at the top, as indicated. At some angle θ_{max}, to be determined, the buckets catch onto a bar

JEAN-VICTOR PONCELET (1788 – 1867)

Born in Metz, in the eastern province of Lorraine, France, a year before the storming of the Bastille brought on the French Revolution, Poncelet would be a soldier for most of his life. Yet in his spare time he made major contributions to mathematics and engineering (mechanics and hydraulics).

Trained at the Ecole Polytechnique, he became a Lieutenant of Engineers. At age 24 he took part in Napoleon's disastrous invasion of Russia, where he was left for dead on the field of battle before being taken prisoner and incarcerated at Saratov, on the Volga. During this period of enforced leisure he wrote an influential mathematical treatise, which helped to lay the foundations of modern projective geometry. It is for this work, published several years later in 1822, that mathematicians chiefly remember him. He was released from prison and returned to France in 1814. A year later, following the collapse of Napoleon's empire, Poncelet returned to Metz as a military engineer. In 1825 he became a professor of mechanics there, and a year later wrote his treatise on undershot waterwheels. With his recommendations the undershot waterwheel, in a modified "breastshot" form, became as efficient as the overshot wheel. The form of Poncelet's waterwheel anticipated modern turbine hydropower generators.

In 1834 Poncelet became a member of the French Academy of Sciences, and a year later moved from Metz to Paris, to take up an appointment as professor in the science faculty at the Sorbonne. He then took charge of the Ecole Polytechnique, with the rank of general, a post he held until the political turmoil of 1848, when he was retired after refusing to serve the Second Empire. At the Crystal Palace Exhibition of 1851, in London, he wrote a report on the English tools and machinery on exhibit. In his later years, Poncelet's interests returned to mathematics, and he wrote numerous articles and a second edition of his projective geometry text.

Compare the two most influential contributors to waterwheel development. Smeaton did make significant theoretical contributions but is best remembered for his experimental, empirical studies. Poncelet's analyses of practical machinery arose out of his theoretical, mathematical training. The fact that both men made telling contributions shows that both the empirical and the theoretical approach to analysis can lead to significant advances.

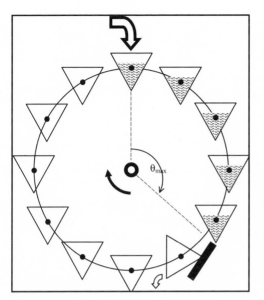

FIG. 2.3. *Idealized overshot waterwheel powered by gravitational potential energy (water head). Water drops vertically from a spout (top) into buckets, and remains there until tipped out at angle θ_{max}.*

that causes the buckets to tip out their load of water. I assume that the buckets do not tip until the bar tips them, so that no water is spilled between angles 0 and θ_{max}—this turns out to be a very convenient simplification.

It is technically possible to build a waterwheel like this, and we can see why it might be plausible to do so. The torque provided by the weight of water, between angles $\theta = 0$ and $\theta = \theta_{max}$, causes the wheel to turn. We denote this type of torque as *gravitational torque*. Such a wheel might work, but it is not the way real waterwheels were built, for practical reasons. First, it would cost more to build this, than a more normal overshot wheel (and where would a medieval millwright get frictionless bearings from?). Second, it does not use one source of available torque (here called *flow torque*) that arises from the water horizontal speed.

This simple model allows us to investigate the influence of gravitational torque separately from that of flow torque. Also it is mathematically unambiguous because we do not need to worry about water leaking from the buckets, or spilling out due to wheel motion, or slopping over the edge into a lower bucket, or a myriad of other annoying realities that might cause the water level to vary with bucket position. (Another unrealistic idealization that I have put into the bucket model is this: the buckets tip effortlessly, i.e., it takes no energy to spill out the water.) I will address these realities later, but for now let us see what the simple model gives us.

First, we find by applying Newton's Laws that a steady, stable wheel rotation does occur. The frequency of rotation is:

$$F = \frac{\rho g f R}{2\pi G_L}(1 - \cos\theta_{max})$$

Here ρ is water density (mass per unit volume, say kg m^{-3}), g is the constant acceleration due to gravity (meters per second per second, m s^{-2}) at the earth's surface, f is water flow rate (cubic meters per second, m^3 s^{-1}), R is the wheel radius (meters, m), and G_L is the *load torque*. Load torque is the useful twisting force of the waterwheel, for example, the torque utilized to turn a grindstone. We want load torque to be large, so the waterwheel can do a lot of work for us. So our simple math model tells us (and this feature carries over into more complicated and realistic models) that low wheel rotation rate is desirable. Another reason for wanting low rotation rate is that we need the water to fill the buckets—if the wheel rotated too fast then the buckets might fly past the spout and not have time to fill up. Yet another reason is that very high rotation rates mean that water may fly out of the buckets because of centrifugal force—and we don't want that.

To determine the efficiency of this simple bucket wheel we need to calculate the power P_{in} that is input to the wheel, by the falling water, and the useful power P_{out} that is obtained. Then wheel efficiency ε is determined by the ratio $\varepsilon = P_{out}/P_{in}$. Consulting once more with Sir Isaac Newton we again find that our simple bucket wheel model gives us a straightforward answer:

$$\varepsilon = \frac{1 - \cos\theta_{max}}{2 + \dfrac{v^2}{2gR}}$$

Here v is water speed (meters per second, m s^{-1}), which is closely related to flow rate f. If the headrace (the sluice that carries water to the wheel) has a cross-sectional area of A, then $f = vA$. The equation for efficiency reveals many important features of waterwheel operation, which we can use to maximize efficiency. For example, efficiency falls as water speed increases. Now water speed is something that the millwright can to some extent control. He can build a headrace with a large cross section, so that water speed is low, for a given flow rate. We see that efficiency increases as wheel radius increases, so the millwright should build his wheel as big as possible. In fact real overshot waterwheels did have big wheels. Last, our equation tells us that efficiency is a maximum for $\theta_{max} = 180°$. Looking at figure 2.3 you can see that this makes sense. Efficiency is maximized when the water tips out at the bottom, because this gives maximum torque.

Let's put in some numbers to see what kind of efficiency the bucket wheel can attain. A reasonable value for water speed is $v = 2$ m s^{-1}, wheel radius is, say, $R = 2$ m, and let us choose the optimum value of 180° for the tipping angle. This gives an efficiency of 95%, which is larger than that attained in reality. We know that this overestimation is because of the idealizations of the bucket wheel model, and so I will now address these issues to form a more realistic model. The qualitative features of the simpler model persist, however: wheel efficiency increases as radius increases, and so on, and so the exercise of constructing an idealized model has provided insight. That is why physicists make use of seemingly abstract idealizations—we saw this earlier with the longbow—a well chosen simple model can capture enough of the real dynamics of a more complicated system to yield insight.

Get Real

I WILL BUILD UPON the bucket waterwheel model to make it more realistic. First, add friction. This takes the form of another torque (yes, *friction torque*), assumed to act about the waterwheel axle. Second, allow for water flow. This is done in figure 2.4, where you will note that the water impinges on the wheel horizontally, or nearly so, as happens most often with real waterwheels. The water flow rate can now contribute yet another torque (*flow torque*), due to water horizontal speed. We therefore have four torques: gravity and flow torques act in the same direction, acting to turn the wheel of figure 2.4 clockwise. This motion is opposed by the load torque and friction torque. Third, and most importantly, we get rid of the buckets, and replace them with realistic *vanes*. Two types of waterwheel vanes are suggested in figure 2.4. These additions make the mathematical model much more realistic.

There is a problem, however. The variety of vanes found in real waterwheels is endless. They have different shapes, widths, depths, and vane an-

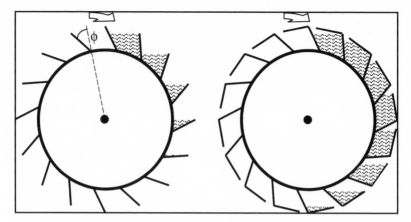

FIG. 2.4. *Overshot waterwheels with two types of canted vanes (vane angle φ). Gravity and horizontal water flow both apply torque to the wheels, turning them clockwise. The design on the left is simpler and would be less costly to construct, but the design on the right is better. The boxes on the right each hold more water, and hold it for almost half a rotation (vs. one-quarter rotation for the wheel on the left), so the gravitational torque is greater.*

gles (shown as φ in fig. 2.4). They spill water at different rates, because of their differing shapes and sizes, but also because of factors I alluded to earlier: centrifugal force, leaks, and other "annoying realities." How can we account for all this variability? Any model that attempts to include all these details will be extremely complicated, and will apply only to one particular waterwheel. Another wheel, with different vane parameters, will require a different detailed model. We want to describe mathematically how waterwheels work *in general*, and not get too involved with different vane characteristics of individual wheels, yet we must address this feature if we are to claim that the model is realistic.

There are times when you can no longer postpone going to the dentist, filling out your tax return, washing your dog, or getting to grips with the details of waterwheel vanes. Happily, we can avoid the details of individual waterwheel variation by putting the whole multitude of possibilities into one bag, which I will label $x(\theta)$ and call the *loss factor*. Thus, different waterwheels will have different loss factors. Consider a single box, defined in figure 2.4 by two adjacent vanes, that fills up with water at the top of the wheel (at $\theta = 0$, since θ is measured clockwise from the top, as in fig. 2.3). The loss factor describes what fraction of the original volume of water is left in the box when it is at angle θ. So, $x(0) = 1$, by definition, because the top box has all the water it received when it was at the top, if you see what I mean. On the other hand the bottom box will certainly have let out all the water (see fig. 2.4), so that $x(180°) = 0$. At angles in between 0° and 180° the loss factor is between 1 and 0. The exact value of $x(\theta)$ at, say, θ = 93° depends on individual waterwheel design details. In figure 2.4 the two wheel vane designs clearly have different loss factors. The good news is that we can proceed a long way toward calculating waterwheel efficiency without knowing all these details, that is, without specifying the form of x. (It turns out that other results obtain, quite independently of loss factor. For example, it can be shown that the steady wheel rotation rate increases with water speed, and decreases as wheel radius increases.)

Consulting Sir Isaac again we find that the expression for overshot waterwheel efficiency has changed to[8]:

$$\varepsilon = \frac{X}{2 + \dfrac{v^2}{2gR}}$$

This has the same form as previously, except that the numerator has been replaced by the *integrated loss factor X*, which is closely related to $x(\theta)$, and is readily calculated once the form of $x(\theta)$ is specified. Let me emphasize this point: there is no physics involved in getting from x to X, merely math. To float you over the jungle, I omit the math.

I can show you how this new equation connects with the old efficiency equation, by working out the loss factor for the bucket wheel. You can see that this loss factor is specified as follows: $x(\theta) = 1$, if θ is between 0 and θ_{max}, otherwise $x(\theta) = 0$. Now putting this into the mathematical mill, cranking the handle and calculating X gives $X = 1 - \cos \theta_{max}$, so that the two efficiency equations agree. For the "canted vane" design of figure 2.4 the X factor can also be calculated explicitly, albeit approximately, if we assume $x(\theta) = 1$ and

8. I have swept a technical detail under the carpet here. If you are interested in such things, you will find them in my technical paper (Denny). Overshot waterwheel efficiency can be less than the value given by the equation, if friction torque and flow torque do not cancel out each other. So here I assume that the millwright built his waterwheel, headrace, and so on, in such a way that these two torques do indeed cancel, in which case the equation is good.

slow wheel rotation. (An efficient wheel will be designed to keep $x(\theta)$ large—as near 1 as possible for as wide a range of θ as possible—to maximize gravitational torque.) For such a wheel

$$\varepsilon = \frac{1 + \sin \phi}{2 + \dfrac{v^2}{2gR}}$$

so efficiency is improved if vane angle is large. Most historical overshot wheels have canted vanes, for this very reason. The choice of ϕ is determined by other factors as well, however. Large ϕ means smaller boxes, and therefore less gravitational torque. So here there is a trade-off—common in engineering design—between competing factors. Gravitational torque wants a small vane angle, whereas efficiency wants a large vane angle. A realistic in-between angle of $\phi = 30°$ (this corresponds pretty well with the vane angles of the wheel shown in fig. 2.5), and the same parameters as before, yields an efficiency of 71%. This is much closer to the value of 63% that Smeaton found by careful measurements, and shows that our model is a reasonable one.

Given different vane designs, we can specify different loss factors and so calculate efficiency for each case individually. I will call a halt at this point, however, as we have uncovered the main influences of waterwheel design parameters on efficiency. Let me summarize them now. An efficient overshot waterwheel will have low water speed, large wheel radius, canted vanes, low wheel rotation rate, and large $x(\theta)$ (low water loss). These are the characteristics of real waterwheels, which suggests that the math model has captured the essential details. I noted earlier that overshot designs improved over the centuries, as millwrights, engineers, and scientists tinkered with the designs. We

FIG. 2.5. *This is the "North Wheel" from a New Jersey ironworks, which was in business around the year 1800 CE. The picture was taken in 1909. Note the water pipe coming in from the left, showing that this is an overshot wheel. The vanes appear to be canted at an angle of about 30°. Photograph by Vernon Royle. Reprinted courtesy Friends of Long Pond Ironworks, Hewitt, NJ.*

can now see *why*, for example, overshot waterwheels grew larger (bigger wheel radius equals more efficient operation). Medieval millwrights didn't know the physics and math, but proceeded by trial and error, keeping the good changes and not repeating the bad ones.

Undershot Waterwheel Efficiency

NOW I TURN TO undershot wheels, and calculate their efficiency. This is a more straightforward calculation than for overshot wheels, if we are satisfied with approximate results. (One reason it is simpler is that there is no gravitational torque for undershot wheels.) I will apply the same analysis to undershot wheels with Poncelet's modifications, and so we will learn why these modifications lead to dramatic improvements in efficiency.

Consider figure 2.6a. This sketches a typical undershot wheel. Note that the vanes are not canted, which is an overshot characteristic of no use here. Water of speed v passes the lower part of our wheel, causing it to rotate. By transferring some momentum to the wheel, the water loses speed, departing the scene with speed v', less than v. Say $v'=cv$ with the constant c somewhere between 0 and 1. Newton's Laws[9] readily yield the output power (Douglas) of the rotating wheel: $P_{out} = \rho A v^3 c(1 - c)^2$. Here A is the area of each undershot wheel vane. Input power (supplied by the flowing water) is $P_{in} = \frac{1}{2}\rho A v^3$ and so efficiency is

$$\varepsilon = 2c(1-c)^2$$

So in my undershot waterwheel model the wheel efficiency depends on the ratio of water speeds, $c = v'/v$, and only on this ratio. It attains a maximum value of $\varepsilon = 30\%$ for $c = 1/3$. This value is pretty close to Smeaton's measurements of 22% for undershots. Of course the ancient millwrights were not knowledgeable about fluid mechanics; they fiddled with vane areas, and so on, until they hit upon the best design, which we now know uses two thirds of the water momentum.

---

9. Applying Newton's Laws to moving fluids is a specialization that has arisen mostly since Newton's time, and has grown during the past three centuries. It is still a field of active interest, and there are several introductory textbooks that illustrate this elegant branch of physics.

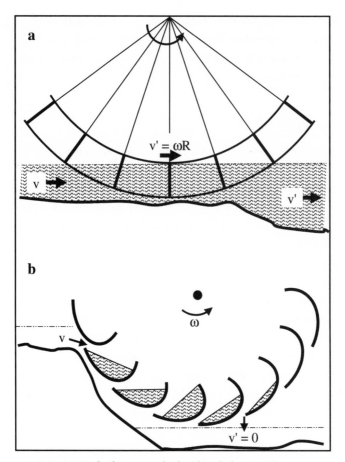

FIG. 2.6. *(a) Undershot waterwheel with radial vanes. The water flow approaching (receding from) the wheel has mean speed* v *(*v'*). The wheel rotates at constant angular speed* v'/R*. (b) Poncelet's modification: curved vanes that trap the water, releasing it only when the water has transferred most of its horizontal momentum. Achieving this requires a careful balancing of vane shape and water flow rate.*

You may very reasonably wonder why the wheel is not more efficient if *all* the water momentum, that is, speed, is transferred to the vanes (so that v' or $c = 0$). After all, this would increase the force F on the vanes, and hence increase the flow torque. Indeed it does, but efficiency depends upon power, not force. The force causes the wheel to rotate, at the speed v' of the "spent" water, so if there is a force then there must be a nonzero v'. The output power P_{out} is calculated from the product of F and v'.

Undershot Waterwheel Efficiency: Poncelet's Modifications

MONSIEUR PONCELET recommended changing the shape of the undershot waterwheel vanes, as shown in figure 2.6b. Water enters a vane higher up than usual for a conventional undershot wheel, roughly speaking water enters the wheel at a level abreast of the wheel axle. So, these wheels were called *breastshot*. They make use of half the gravitational energy harnessed by an overshot wheel, while remaining technically undershot, since the water passes underneath the axle.

The key feature to note is the curved vanes. These were carefully shaped to hold the water as the wheel turns. When the vane angle becomes too steep, the water falls back (off the outside edge of the vane, as shown in fig. 2.6b) with zero speed, $v' = 0$. This can only happen if the vane is properly adjusted to the water speed. For example, if the water fell off too soon, then it would not be exerting torque on the wheel as much as it might. On the other hand, if the water flew off the front of the vane, then it would not be transferring all of its momentum to the vane, and again torque is reduced. Note, the vanes are not designed to hold *all* of the water that flows onto them, but are carefully shaped to retain as much as possible, for as long as possible.

When the waterwheel efficiency is carefully recalculated for Poncelet's vanes, taking into account the extra momentum transfer due to the portion of the flow that is retained in the vanes, we find that the efficiency equation has changed to

$$\varepsilon = 2c(1-c)$$

You must anticipate that this expression underestimates the efficiency of the Poncelet wheel, because my calculation does not account for the gravitational torque of breastshot wheels, but includes only the effects of water flow rate through the modified vanes. The new equation yields a peak efficiency of 50%, corresponding to[10] $c = 1/2$. This increase represents a significant im-

---

10. If v' is zero then why is c not zero? Because not all the water flow is held in the vanes. So, we reconcile $v' = 0$ with $c = 1/2$ by saying that half the water transfers all of its momentum to the waterwheel, whereas the other half transfers nothing.

provement, but falls short of Poncelet's figure (65%) for the reason just stated. Nevertheless we have seen that Poncelet's modified undershot wheel, even without the gravitational torque, is significantly more efficient because it has a different dependence on the parameter c. Less energy is removed from the flowing water than with conventional undershot wheels, but the removed energy is more effectively utilized.

What Goes Around, Comes Around, Again

WATERWHEELS WERE OF economic importance well into the twentieth century. They were then eclipsed by turbines—their logical offspring. (See fig. 2.7 for an 1868 picture of all three waterwheel types, and some early turbine designs [Brown].) Today there is a resurgence of interest in these ancient and stately machines. If you look on the web you will find all kinds of advice from enthusiasts about building your own waterwheel. There are many commercial waterwheel builders who will sell you one, if you prefer. In preparing this chapter I found these people to be very helpful (see fig. 2.8). Clearly business is booming. Why this interest? For me, in part, it has to do with the application of science to technology, but this can only be part of the explanation, since technology does not turn everyone's crank, to coin a phrase.[11] Many waterwheel enthusiasts, I think, find that the historical echo resonates within them, going by the large number of waterwheel restoration projects that are happening around the world. There is, in the United States, a society for the preservation of old mills. In both the Old World and the New World, waterwheels were important to our economic and industrial development, to the extent that they have become cultural icons.

Windmill Development

NOW FOR SOMETHING completely different, to paraphrase Monty Python. Well, as different as two brothers can be, for the waterwheel and the windmill are from the same stock and have a lot in common.

---

11. But many people are interested in the technical details, judging by the numbers who have asked me for reprints of my technical article. These are mostly people who want to build their own waterwheels—good for you. One unusual request came from the Spencerville Mill Foundation in Ontario, Canada, who would like to use my paper to help educate the visitors to their mill. Good for them, too.

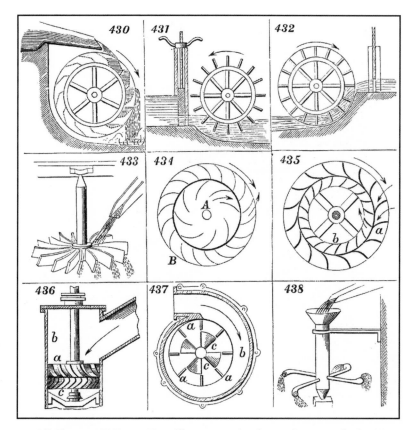

FIG. 2.7. *From an 1868 text. You will now recognize the overshot waterwheel (430, with angled vanes), undershot waterwheel (431), and breastshot waterwheel (432— not the most efficient—it should have curved Poncelet vanes). 434 is an early reaction turbine, designed by Poncelet's compatriot Fourneyron in 1834. Water enters at A and exits at speed through the vanes, causing the outer wheel to turn. 435, 436, and 437 are other turbines. Reproduced with permission of The Astragal Press.*

There is no evidence for the use of windmills in classical Greece or Rome (Landels). This is perhaps surprising, given the engineering skills exhibited by Greeks and Romans in other fields. Maybe the water-powered mills were more reliable; a key feature of windmills is their relatively restricted geography. The Islamic vertical windmill—resembling a revolving door (James), complete with cylindrical housing—was developed in windy Persia or Afghanistan in the ninth or tenth century CE. It was widespread in parts of the Middle East in the tenth century when it was reported by an Arab traveler

FIG. 2.8. *A modern all-metal wheel, being tested prior to delivery. "On our current production, we are building a 35ft Fitz type wheel for a mill restoration, I am in two weeks installing a Poncelet wheel in Vermont"—Robert Vitale. In an age of hydroelectric power stations, people still want waterwheels. Thanks to Bob for this image.*

as a "conspicuous feature" of these parts, used largely for irrigation (Usher). Perhaps the Crusaders brought the idea of windmills back to northern Europe; certainly the first reports of windmills in Germany and Holland date from this period. On the other hand, the European windmills were always of the horizontal kind (Mason), and this distinction may suggest an independent invention. Whatever their origins, windmills were a common feature of flat Holland and the north German plains by the twelfth century CE. At about this time, or shortly thereafter, windmills appeared in the flat fenlands of eastern England, and in northern France (recall Edward III at Crécy, climbing a windmill to survey his French enemy).

For 600 years the windmill would be an important (indeed crucial, in the case of Holland) aspect of life. It would develop and improve, as inventive minds were turned upon it, motivated by a growing need for energy. Of course

FIG. 2.9. *A light twentieth-century metal windmill, with fantail. This one is in Albuquerque, NM. Photograph by the author.*

in the end it fell by the wayside along with its waterwheel brother as steam power arose, and by the twentieth century windmills were pretty much obsolete. They did hang on in certain niche applications, however. For example the light steel-girder windmills, with steel sails, provided power to many prairie farms in the American midwest during the early 1900s (Dresner). These windmills (fig. 2.9) will be familiar to many of you. Many other windmills remain as rural museum pieces, and many countries have societies dedicated to their preservation.

The earliest windmills in Europe were the *post mills*. These consisted of a large pole with rotating sails at the top, and the mill beneath. The whole mill, sails and all, could be rotated about the post as wind direction changed. This was usually achieved by the miller turning a capstan wheel attached to winding gears. Here we have the main reason why windmill development was slower than that of waterwheels, and why wind power was limited geographically. Winds are more variable than water. Water can be channeled so that, by adjusting sluice gates, the headrace flow is maintained constant (except in drought conditions). The four winds, however, can gust and blow, or fall calm.

Wind power can change in a few seconds, and wind direction is beyond the miller's control. So the windmill must rotate to face the wind, whereas the waterwheel had water piped to it.

You will not be surprised to learn that the construction of a building that could be turned on a dime by a single miller was not trivial, and this fact alone is enough to explain the slow utilization of wind power, as compared with water power exploitation. By the 1500s *smock-mills* appeared alongside post mills. These mills had fixed wooden frames, hanging like a peasant's smock, over the mill body. Only the cap at the top of the mill, and the sails, turned about the post. This development was significant because the weight of material to be rotated about this vertical shaft was reduced. Reduced weight meant that the sails could be turned more quickly in response to a change in wind direction. Conversely, bigger sails could be constructed, so smock-mill power exceeded that of the older post mills of the same size. This process was extended when the *tower mills* (or turret mills) appeared. These consisted of stone or brick bodies, with only the cap and sails rotating about a vertical shaft. The stronger, taller bodies of tower mills meant that even bigger sails could be attached. The diameter of sails was now limited (to about 90 feet) by the available lengths of pitch-pine timber spines (*stocks*) rather than by windmill height. These tower mills could power several sets of millstones, and by the seventeenth century they were the backbone of Dutch life. The tower mills are the type we usually first think of nowadays, since many of them have survived to the present day quite well—being younger and more strongly built. They represent the pinnacle of windmill development. The windmills shown in figure 2.10 are either smock or tower mills. The complicated gearing of an industrial mill is shown in figure 2.11; this picture gives us a feeling for the size and complexity of large windmills (and watermills).

Apart from developing in form, many bells and whistles have been added to windmills through the years. A wooden friction brake acted on the drive wheel, preventing the mill from damage during periods of high winds. Fantails (like the tail rotors of helicopters; see fig. 2.9) were added to automatically rotate the mill, so it always faced the wind. Another sixteenth-century introduction was the inclined sail. An example of this is shown in figure 2.12. It may seem surprising, but the sails were most efficient when the axle about which they turned (the *wind shaft*) was inclined slightly (about 10°) to the horizontal. This is because, we now know, the wind is slowed down close to the ground, by friction with the surface, and so a typical "velocity profile" shows

FIG. 2.10. *Windmills were a vital aspect of Dutch industry and life for several centuries, and were used to drain lakes and reclaim land. A thousand of them can be seen across the flat landscape. I am grateful to the Dutch village of Kinderdijk for permission to reproduce this picture from their Web site.*

wind speed increasing with height above the ground. This velocity profile induces a slight downward component to the otherwise horizontal wind force. (Think of a tower of toy bricks toppling. The upper bricks develop higher speeds as the tower falls. Here it is the other way round: the wind falls because of the higher speeds at the top.) So, to catch the wind better, the sails are inclined. Sails themselves evolved a great deal over the centuries. The pitch angle of an individual sail[12]—the angle it presents to the wind—is greater near the axle than further out, just as for an aircraft propeller. (Needless to say the millwrights who first introduced these developments were working empirically—they knew nothing of aerodynamics.)

Sails originally were dressed with sailcloth. These sails could be trimmed in strong winds (10 m s^{-1}) or furled in very strong winds (exceeding 12 m s^{-1}). In 1772 hinged shutter sails were introduced in England, and spread to

---

12. There were usually four sails per windmill, as you can see in figure 2.10.

FIG. 2.11. *The complications of windmill design led to slower development of wind-powered generators than of water-powered generators. I am grateful to De Hollandsche Molen for permission to reproduce this old photograph of the complex wooden and iron gearing inside an industrial windmill (A) and to Vernon Maldoom for the picture of whirring gears in a working watermill (B).*

Holland. These shutters resembled Venetian blinds; they automatically adjusted their pitch as wind speed increased, due to wind pressure, and this adjustment provided a more or less constant operating speed. Such automatic adjustment to changing conditions is nowadays called *feedback control* (windmills provide a number of early examples). I discuss feedback control in detail, in a later chapter. Airbrakes were an 1860 English innovation that effectively slowed the sails in high winds. The fact that windmills were evolving even during the later stages of the first industrial revolution shows that they remained economically important, at least in rural situations.

The Dutch involvement with, affection for, and development of windmills arose because of the vital nature of these machines—vital for all aspects of Dutch life for several hundred years.[13] Holland owes its physical creation as well as its economic development to windmills, since the application of wind power to pump water. From 1414 the land of Holland was beginning to be reclaimed from the sea. Sea defenses were being built and lakes and marshes were being drained. Windmills pumped the excess water into canals. This ex-

---

13. The Dutch Windmill Society maintains an excellent Web site: http://webserv
.nhl.nl/~smits/windmill.htm.

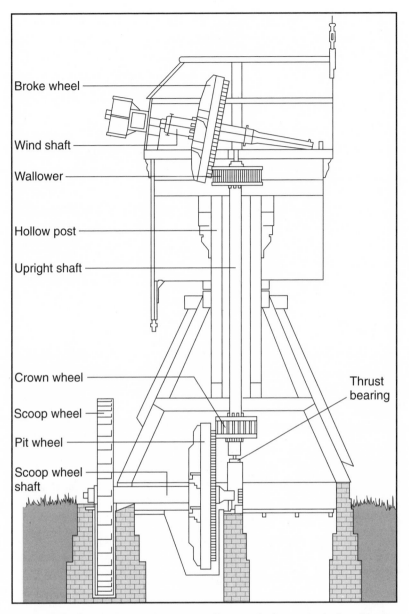

Broke wheel

Wind shaft

Wallower

Hollow post

Upright shaft

Crown wheel

Thrust
bearing

Scoop wheel

Pit wheel

Scoop wheel
shaft

FIG. 2.12. *A drainage pump—this is essentially an undershot waterwheel working in reverse. Note the inclined axle (wind shaft). I am grateful to De Hollandsche Molen for permission to reproduce this diagram.*

tensive and continuous pumping freed land for agricultural use. The flatness of the land was responsible for the sea encroachment, and the vulnerability of the population to flooding, but it also provided the means to pump water, since winds were uninterrupted by topographical features.[14] The wind-powered water pump (fig. 2.12) acted in the opposite direction to a breastshot waterwheel. Water was pumped uphill using external (wind) power to turn a scoop wheel. Contrast this with waterwheels, where falling water provides power to turn a wheel. The maximum lift of these water pumps was little more than 4 feet, limited by the wheel size, so larger lift heights required many pumps working in a series.

Apart from drainage, the Dutch milling applications extended from grinding flour and sawing wood to milling paper, pepper, and snuff, to dyeing, fulling, and tanning in the cloth industry. The importance of windmills may be judged by the fact that there were approximately 9,000 of them in Holland in the nineteenth century, about one mill per thousand of the population.

Windmill Efficiency

BECAUSE OF FRICTION in the complicated gear train, and other sources of inefficiency (such as the sails not facing directly into the wind) early windmills could not operate in light winds, less than 5 or 6 m s^{-1} (11 to 13 mph). At these speeds the sails would just turn, but could do little work. At the other extreme, we have seen that the sails would have to be trimmed if wind speeds exceeded 10 m s^{-1} to avoid damage. Thus the range of usable wind speeds was limited, and in practice this meant, in Holland, that a windmill of the older type could operate for about 2,700 hours per year (say 7 hours per day) on average. Later, more efficient windmills could operate in winds exceeding about 4 m s^{-1} and this made quite a difference. Because such light winds are very common in northern Europe, the Dutch windmill could now operate for an average of 4,400 hours per year (12 hours per day). So, a straightforward increase in efficiency permitted a vast increase in productivity. Also, the lower minimum wind speed meant that the later designs produced more useful power, in a given wind, than did the earlier machines. At the peak of their development, windmills could sustain 50 horsepower output (37 kW), and in-

---

14. Dutch landowners prohibited the planting of trees, so that their mills would have free air to power them.

termittently could double this power. They were thus more powerful than the best waterwheels. But how efficient were they?

Happily I do not need to do any work to tell you about windmill efficiency because this calculation has already been done (Massey). In the notation of my undershot waterwheel efficiency calculation, the windmill efficiency has been found to be

$$\varepsilon = \frac{1}{2}(1+c)(1-c^2)$$

In the calculation it has been assumed that no energy is lost through the gears, or anywhere else. Efficiency is defined as the fraction of wind power that is taken from the wind by the sails. So, if the wind power entering the area swept by the sails ($A = \pi R^2$ for sails of radius R) is P, then the wind power exiting this area is $(1 - \varepsilon)P$. From the equation we see that the maximum power occurs for $c = 1/3$, as with undershot waterwheels, and this yields a maximum theoretical efficiency of $16/27$ or about 60%.

If only. In practice windmills are nowhere near this efficient. Because of the complicated gear train, imperfect sail alignment, sail mass (and cap mass if the wind is changing direction), and a host of other reasons, real windmill efficiencies are typically more like 5%. This fact renders the theoretical calculation more or less meaningless, which is why I have treated it rather cursorily.

Let us see what kind of output power we get for a 5% efficient mill. Take a bunch of wind with mass m; the energy of that wind is $1/2\, mv^2$, where v is wind speed. The power of the wind flowing through the sail area is therefore $1/2 f v^2$, where $f = \rho v A$ is the wind flow (mass per unit time through the area A). Recall that ρ is air density. This calculation gives us a wind power of $P = \frac{\pi}{2}\rho R^2 v^3$. What is important here is that wind power increases fast as sail radius increases, and even faster as wind speed picks up. A 10-meter sail operating in a 10 m s^{-1} wind will generate 32 times more power than a 5-meter sail operating in a 5 m s^{-1} wind. So, design your windmills to have large sails that can work in strong(ish) winds. Lo and behold, that is the way windmills developed historically.

Putting in typical numbers ($\rho = 1.2$ kg m^{-3}, $R = 10$ m, $v = 5$ or 10 m s^{-1}) with 5% efficiency gives us an operating windmill power of between 1.2 and 9.4 kW (2–12 horsepower). This is right in the middle of the figures measured for moderate-sized windmills. Thus you see that windmill output power is significant, compared with that of a waterwheel, even though the windmills are so inefficient. I guess that efficiency is not the crucial factor in windmill design, given that wind is handed out free. Construction costs, rather than efficiency, must be the driving factor for windmills.

REFERENCES

Britannica® CD 98 Standard Edition. *Waterwheels.*

Brown, H. T. (1990). *Five Hundred and Seven Mechanical Movements,* p. 104. Morristown, NJ: The Astragal Press.

Coriolis G. G. J. (1835). *l'Ecole Polytechnique* 15:142–144.

Denny, M. (2004). *European Journal of Physics* 25:193–202.

Douglas, J. F., and R. D. Matthews (1996). *Fluid Mechanics,* chap. 8. Harlow, United Kingdom: Longman.

Dresner, D., ed. (1998). *The Hutchinson Encyclopedia.* Godalming, United Kingdom: Helicon.

Gies, F., and J. Gies (2002).*Cathedral, Forge and Waterwheel: Technology and Invention in the Middle Ages.* New York: HarperCollins.

James, P., and N. Thorpe (1994). *Ancient Inventions,* chap. 9. New York: Ballantine.

Landels, J. G. (1978). *Engineering in the Ancient World.* London: Constable.

Magnusson, R. J. (2001). *Water Technology in the Middle Ages,* p. 100. Baltimore: Johns Hopkins University Press.

Mason, St. F. (1962). *A History of the Sciences.* New York: Collier Books.

Massey, B. S. (1997). *Mechanics of Fluids,* p. 131. London: Chapman & Hall.

Poncelet biographical sketch, sources: O'Connor, J. J., and E. F. Robertson, article on the University of St. Andrews mathematics department Web site, available at: http://www-groups.dcs.st-and.ac.uk/~history/BiogIndex.html; "Jean Victor Poncelet," available at: http://96.1911encyclopedia.org/P/PO/PONCELET_JEAN _VICTOR.htm; Lycée Poncelet de Saint-Avold Web site, www.ac-nancy-metz.fr/ Pres-etab/PonceletSaintAvold/Histoire/JVP.htm.

Smeaton, J. (1759). *Philosophical Transactions of the Royal Society* 51:100–174.

Smeaton biographical sketch, sources: Uglow *Lunar Men* (see below); Institute of Civil Engineers, www.icivilengineer.com/Famous_Engineers/Smeaton/; American Society of Civil Engineers, on history and heritage, www.asce.org/history/ bio_smeaton.html; JSTOR Library Services of University of Michigan; Wikipedia, http://en.wikipedia.org/wiki/John_Smeaton.

Uglow, J. (2002). *The Lunar Men.* New York: Farrar, Straus & Giroux.

Usher, A. P. (1954). *A History of Mechanical Inventions,* chap. 7. New York: Dover.

THREE

COUNTERPOISE

SIEGE ENGINES

Etymology

INGENIUM IS MEDIEVAL English vernacular for "an ingenious contrivance." In those days the word was used to describe siege engines, those weapons of war designed and widely employed to reduce castles to rubble. The crew (and these engines from the Middle Ages required a *large* crew) were *ingeniators*—hence, our modern "engineers." All the machines in this book are well described as ingenious contrivances, and so I have included this strange word in the title. These machines in their various forms say so much about us, and none more so than the medieval version of the siege engine.

There is a linguistic progression here that mirrors the diversity and the development of siege engines. We have already encountered the *euthytonos* and *palintonos*, large catapults from classical Greece. The Romans—no laggards when it came to military engineering—contributed the *onager*, among others (Landels). These western European creatures, however, were destined for extinction. The Chinese, a source of many innovations that are all-too-often taken to be of occidental origin, employed *traction trebuchets* at about the same time that the onager stalked the earth. The oriental beasts migrated westward, evolving all the time, diversifying and, above all, growing into behemoths. They differed from the classical European catapults in their source of power. Their projectiles were powered by gravity rather than by twisted rope or sinew, or by stretched metal springs. This class of siege engine is referred to as *counterpoise*. Counterpoise engines took many forms, and were given many different names in different languages (Chevedden 2000), including: *bricolle, pierriere, petrarie, couillard*,[1] *biffa, matafunda, robinet,* and *tormentum*. The Arabs, who contributed significantly to the development and spread of the larger versions, called one creation "daughter of the earthquake," an evocative moniker that should send you a wake-up call about the scale of these weapons. I am not talking one-ounce arrows here.

To persist with the zoological metaphor, the herd of counterpoise beasts, firmly established in Europe by 1200 CE, diversified into a veritable menagerie, as odd as the names given to them. They grew large, and the two largest are the ones of interest to us here. These are the *mangonel* and, at the top of the food chain (the *Tyrannosaurus rex* of the siege-engine world), the medieval European *counterpoise trebuchet*. The last examples of original medieval trebuchets were destroyed in a fire in Arras, France, in 1830. Figure 3.1 shows a modern Danish reconstruction of a full-sized machine. We will dissect these leviathans later in this chapter.

Popular and Enduring

TREBUCHETS WERE LAST built (for use in war—an important qualification) in the New World at the beginning of the sixteenth century CE. Cortés had one such machine constructed to assist his pacification of the Aztecs. Sadly for him, though perhaps not for the Aztecs, the knowledge of trebuchet construction was fading throughout the world because of the advent of gunpowder weapons and, in particular, of larger cannon. These artillery weapons re-

---

1. I will not translate this Old French word, lest my readers' cheeks be easily set aflame by indelicate anatomical references. Just look at some of the pictures, and you will get the idea.

FIG. 3.1. *(A) A modern reconstruction of a large medieval trebuchet siege engine. Note the massive hinged counterweight attached to the short lever arm. Soldiers on the left provide a sense of scale. The large traction wheels are for raising the counterweight. (B) The trebuchet being loaded. You can see one end of the sling attached to a hook at the end of the long lever arm. Thanks to The Medieval Centre, Nykoebing F, Denmark, for providing these images.*

placed trebuchets at about this time in history, after the counterpoise engines had ruled the known world for three centuries (Keen). The Spanish engineers employed by Cortés did not get it right the first time, and there would be no second chance. The first shot fired by Cortés' engine flew straight upward and then, gravity being what it is, fell straight downward, onto the engine, smashing it to pieces (Castillo).

Despite this ignominious end to their official career, after an interregnum of four centuries, the trebuchets continue to be built, admired, written about, analyzed, and invoked.[2] They really do say a lot about us humans:

• *Cruelty.* During many medieval sieges, the bodies of dead soldiers or rotting animals were shot over the walls of a besieged castle to spread disease—perhaps the first example of biological warfare. For example, during the Mongol siege of Kaffa, in the Crimea in 1345–1346 CE, plague-infected corpses were flung over the city walls. Escaping Genoese merchants carried the disease to Western Europe, where it exploded as the Black Death (Chevedden).

• *Eccentricity.* A present day English landowner, Hew Kennedy, has constructed a large trebuchet on his property, just for fun. He launched a variety of projectiles from it: a 112 lb (50 kg) iron weight that flew 235 yards (215 m), a large dead pig (160 m), a one-ton automobile (80 m), upright and grand pianos (135 m and 115 m, respectively), and a toilet filled with gasoline, set alight (range unrecorded).

• *Humor.* Newspaper headlines for the last-named projectile: "Those Magnificent Men and Their Flaming Latrines" (Hadingham; Mapes).

• *Inventiveness.* Applied ingenuity is one of the main topics of this book. Read on.

• *Education.* Trebuchets are useful machines for teaching physics and engineering principles to undergraduate students. If you search the Web you will find many university physics and engineering departments, in particular, in the United States, that have instigated student projects to build these engines. The math analysis can be as simple or as complicated as you like, and the basic idea is easy to visualize, because to a good approximation the only force involved is gravity.

• *Inquisitiveness.* In addition to formal education, siege engines, in gen-

---

2. Recently, in the blockbuster fantasy film *Lord of the Rings*.

eral, and trebuchets, in particular, satisfy the need of many people to tinker with machines, and to learn about our past. In Europe many museums and historical societies (in Denmark, Sweden, England, Scotland, and in particular, France) exist where full-scale siege engines have been reconstructed and fired. Consequently much has been learned about the construction and operation of the medieval machines. Some of the pictures in this chapter come from these reconstructions. Many hobbyists in the Untied States build smaller-scale machines. These machines are available commercially, along with trebuchet simulators, and are often backed up with detailed technical analyses.

• *Entertainment.* There have been several TV documentaries and reconstructions of siege engines, for our education and entertainment. A NOVA program originally broadcast on PBS in February 2000 brought together fifty craftsmen from England, Germany, the United States, and France (including Renaud Beffeyte, a master carpenter and the world's only full-time professional medieval trebuchet builder) on the shores of Loch Ness, in Scotland. They built and fired a large machine, demonstrating the accuracy and destructive power of the mangonel (NOVA).

• *All of the above.* In his determined attempt to conquer Scotland, the fourteenth-century English King Edward I (Hammer of the Scots and grandfather of Edward III, whom we met at Crécy) besieged Stirling Castle. Irritated by the unwillingness of the besieged to be conquered, he ordered the construction of a massive trebuchet, which came to be dubbed War Wolf. The Scots, seeing this monster being assembled outside their walls, decided after all that it would be better to surrender. Edward, by this time curious to see how War Wolf would perform, declined their offer.

Perhaps the main human quality omitted from this list is romance. It is a little difficult to associate tender feelings with a 20-ton giant war machine (Gravett). Maybe I can try. Imagine, after the siege of Stirling, an English engineer in Edward's army meets a Scottish girl. Engineer: "Oh Fiona, the way your golden tresses contrast with the smoldering ruins of your shattered city touches my heart." Girl: "Oh Rupert, I am moved at the sight of your magnificent hinged counterweight." Hmm, maybe not.

Technical Development

THE DIFFERENCE BETWEEN the siege engines of classical Europe and the early counterpoise engines of China, later of medieval Eurasia, is the power source. Torsion engines survived in Europe until the early medieval period, at which time the superior gravity-powered counterpoise engines replaced them. All counterpoise engines are based on the lever principal. A large (and sometimes enormous) counterweight is attached to the short arm of the lever, as in figure 3.1. The projectile is attached to the long arm, either directly or via a sling. Raise the counterweight, let it go, and gravity will do the rest. That is the simple principal. As so often happens with technology, however, the devil is in the details.

The first traction trebuchets appeared in the fifth through third centuries BCE in China. Here *traction* refers to the fact that these early machines were powered, not by gravity, but by soldiers, as many as 250 of them to each machine. These soldiers would, at a given signal, haul on ropes attached to the short end of the lever, and so propel the projectile, which was usually a stone sphere. This method has a few obvious disadvantages. First, the torque applied by the soldiers would not be constant from one shot to the next, and so maintaining consistent range would be difficult. Second, a closely packed group of 250 soldiers would be a relatively easy target for enemy archers.

By about 600 CE counterpoise engines, now with counterweights, had spread westward across Asia and had reached the Arab world in the Middle East and North Africa, and the Byzantine empire of eastern Europe and Anatolia. From here the counterpoise engines spread to Europe, where they flourished and grew. The earlier of the two large forms that I will consider is the *mangonel*. I am here following common but not universal practice in nomenclature: trebuchets are counterpoise engines with hinged counterweights, whereas mangonels are counterpoise engines with fixed counterweights.

Whether mangonel or trebuchet, the counterpoise siege engines had two significant advantages over the older torsion engines, and these advantages fully account for the demise of the torsion engines in the early Middle Ages. First, the counterpoise engine action was smooth. The lever arm (the *beam*) accelerates rapidly, releases the projectile, and then decelerates, all in one smooth continuous motion. Such smooth action is not the case for torsion engines. As we saw in the first chapter, the torsion engine throwing arm is stopped suddenly by a padded bar (recall fig. 1.9). This bar or stop brings large forces into play that can break the engine. There is an equally important con-

FIG. 3.2. *The trebuchet of figure 3.1 launching a smoke bomb. Thanks to The Medieval Centre, Nykoebing F, Denmark, for providing this picture.*

sequence of smooth versus jerky action, and this has to do with the siege engine target. To destroy a stone castle tower or curtain wall, it was necessary to fire numerous projectiles at one location, pounding it again and again. But the jerky torsion engine frame would shift after each shot, as the throwing arm slammed into the padded bar when the projectile was released.[3] So the

---

3. The Roman onager torsion engine was so called because onager means wild ass, because of the kicking motion. Glance at figure 1.9B and you will have no difficulty in visualizing this effect.

engine frame would have to be reset after each shot. This resetting action considerably slowed down the siege engine rate of fire, and also significantly degraded accuracy because the engine was not always repositioned exactly. For the counterpoise engines, the smooth action meant great consistency in accuracy, since there was no kicking and no need to reposition. The modern reconstructions of medieval counterpoise engines confirm that shots could be placed with consistency into a small area. (A Danish engine has been able to group a large number of shots in a 6-m square, at a range of 180 m [Hansen].) Thus the counterpoise engines were more accurate and had a greater rate of fire than torsion engines.

The second advantage of the counterpoise engine is that it could grow big—there is no fundamental limitation placed on its size. This is not the case for torsion engines, because of the large forces involved when the throwing arm is stopped. So, in the early Middle Ages castle builders found that they were challenged by accurate counterpoise engines with a relatively rapid rate of fire (perhaps once every half hour), and that these engines were growing ever larger. Worse, the counterpoise engine range could be greater than the range of an arrow. This is important because archers were a good defense against siege engines. From the castle battlements, they could send a shower of arrows against the engineers who were operating the machines. The engineers would necessarily be huddled together and must have presented an easy target, especially as they were preoccupied with working the engine and not looking skyward to avoid arrows. But defensive archers became less effective as the siege engine range increased.

So how did the castle builders respond to the threat posed by large mangonels and trebuchets? They built larger castles with much thicker and stronger towers and walls. Outer walls and inner walls. Massive ramparts. This response to the counterpoise engine threat is an early example of an arms race: castles became bigger to counter bigger siege engines, which themselves became bigger to knock down the bigger castles, and so on. The first large castles encountered by western Europeans were those that they saw in the Middle East during the Crusades. The crusaders must have been completely awed by the likes of Crac des Chevaliers (Castle of the Knights—fig. 3.3) in Syria, a monumental construction that was much larger and better built than any castle they had seen back home. Of course, this superior castle construction reflects the fact that counterpoise engines arrived in the Middle

FIG. 3.3. *Crac des Chevaliers, Syria. One of the most imposing fortifications ever built. The strength of the walls of this crusader castle is evident. The massive scale may be judged by the figures on the tower (just above the skyline). The strength and size of this well-preserved castle (the outer walls, 30 m thick, were built more than 800 years ago) testify to the effectiveness of large counterweight siege engines. (A) Photo by M. Disdero. (B) Photo by Frans Dekker www.fransdekkers.nl/: the revetted inner walls (reproduced with permission).*

East from the orient. Subsequently, the crusaders took home with them the idea of bigger and better siege engines and larger castles.

The range of a mangonel or trebuchet depended on the weight of projectile it fired. These projectiles were usually stone spheres, with a weight up to one ton (Chevedden), but more commonly around 20–100 kg (say 40–220 pounds). Medieval engines were claimed to have ranges of 150–350 m (Hansen; Payne-Gallwey). Occasionally, as I mentioned earlier, more bizarre projectiles would be fired. Sometimes live spies, as well as dead soldiers, would be seen flying in a graceful arc across the sky as they headed over a castle wall. It is said that the engineers knew when a live projectile had arrived at the target when the screaming stopped[4] (Hansen).

Taking Engines Apart

I BEGIN THE TECHNICAL analysis of counterpoise engines in the next section. First, though, I need to say what is to be calculated. There are two measures of engine effectiveness that I will explore: range and efficiency. Range is fairly straightforward. Compared with arrows, the aerodynamics of stone spheres is simple.[5] I have already indicated the importance of siege engine range for performance, but why does efficiency matter? Apart from the obvious reason that higher efficiency means longer range, all else being equal, there is a quite separate reason connected to the engine recoil following projectile release. I said earlier that counterpoise engine action was smooth—no sudden jerks—which is good, but has the deleterious consequence of a long settling time. That is to say, the engine beam, postrelease, rocks back and forth for quite a while. Until the beam motion stops, the engine cannot be reloaded, so long recoil time adversely influences rate of fire. It is not all that easy to stop a 10- or 20-ton weight from rocking, if it wants to rock, and so it is a good idea to make the engine as efficient as possible. A more efficient engine means less energy available for the counterweight to rock, since more of the energy has been taken away by the departing projectile. A sensible definition

---

4. Perhaps so, though this seems a little implausible to me. Given the accelerations involved, I would expect the screaming to stop during the launch phase.

5. Range calculations for screaming spies, rotting pigs, and flaming latrines will be set aside. If any reason is needed for this, then let us say that there are insufficient data on their aerodynamic drag coefficients.

of efficiency for a counterpoise engine is the ratio of projectile energy to input energy. Projectile energy is the energy due to the projectile launch speed, and input energy is the energy required to raise the counterweight to its launch position. Greater efficiency means more energy goes to the projectile and less is left behind in the frame, and so less energy is available for recoil. Thus, greater efficiency means faster rate of fire, as well as longer range.

Several design features must be analyzed if we are to understand what makes a mangonel or a trebuchet tick. We will see that the trebuchet is the better machine, with longer range and greater efficiency. Why should the existence of a hinged, rather than fixed, counterweight make a better machine (apart from impressing Fiona of Stirling)? And then we need to consider the influence of the projectile sling.

Some of the classical period European torsion engines had slings, and some did not. All the medieval counterpoise engines had slings. For machines without slings, the projectile sat in a spoon-shaped depression at the end of the long lever arm of the main beam. For machines with slings, the projectile sat in a pouch that was attached to a strong rope, the other end of which was attached to the main beam. The ratio of sling length to long and short lever arm lengths was crucial.

To give you an idea of the various features that I will be investigating, consider the diagrams of figure 3.4. The main beam consists of a short lever arm (length, ℓ) and long lever arm (length, L), on either side of the main pivot. The counterweight (mass, M) is attached directly to the end of the short lever arm, for the mangonel. For the trebuchet, a hinge of length h connects counterweight and short lever arm. The sling length is r.

Figure 3.4 brings out another feature of mangonel and trebuchet design. Sometimes a slit trench was dug beneath the engine, so that the projectile pouch would not drag along the ground prior to becoming airborne. Also, the projectile would be given a "gravitational assist" at the beginning of its trajectory prior to release, picking up speed before being slung round and up. Mostly, however, slit trenches were not dug; instead, a horizontal slide or trough was constructed at the bottom of the engine. The projectile, in the sling pouch, would slide along rails until it became airborne. The slide was preferred over the trench for two reasons. First, if a different target were chosen, say a different part of the castle wall, then our siege engine would have to be reaimed and, without a slide, reaiming would mean digging a new trench. So, trenches were dug only if the engineers anticipated a lot of shots at the same

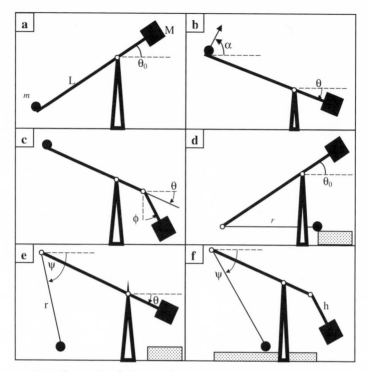

FIG. 3.4. *Siege engine designs. (a, b) Case I. The basic lever arm engine. (c) Case II. Add a hinged counterweight. (d, e) Case III. The mangonel: fixed counterweight, plus sling. (f) Case IV. The trebuchet: hinged counterweight plus sling. The trebuchet projectile is here constrained (it slides along a rail before liftoff), whereas the mangonel projectile is here unconstrained (it can swing low). In all cases m is projectile, M is counterweight, $L+\ell$ is beam length, pivots are represented by open circles, beam angle θ has an initial value $-\theta_0$, h is counterweight hinge length, r is sling length, ϕ is hinge angle, ψ is sling angle, and α is launch angle.*

target. Another, perhaps more important, reason is that slides permitted longer slings. (A medieval trebuchet, with long sling, is shown in fig. 3.5.) I will show you that longer slings gave longer range. A long sling would require a deep trench (maybe 20 feet), which is unrealistic.

Four counterpoise siege engines are shown in figure 3.4. Only two of these existed in reality, the mangonel (beam plus sling, with fixed counterweight) and trebuchet (beam plus sling, with hinged counterweight). The other two machines are hypothetical, and are included so that we can gain understand-

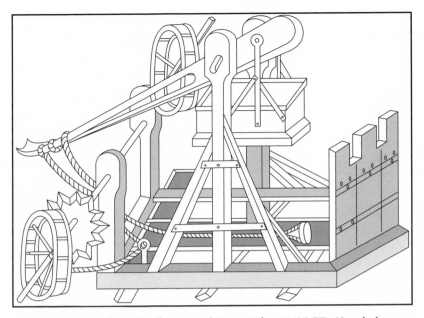

FIG. 3.5. *Medieval illustration from central Europe (about 1405 CE). Note the long sling and hinged counterweight. The wheels on the left are probably part of a treadmill winch for raising the counterweight. The shield in front suggests that the designer was not confident his machine had a range exceeding that of enemy arrows.*

ing about the beneficial influences of hinge and of sling. The first hypothetical engine is the simplest possible counterpoise engine, consisting simply of the beam. No sling, and no hinge. I will label this no-brainer lever-arm engine case I. It has a couple of virtues for us, though none at all for a medieval engineer. (He would be better off with an old torsion engine rather than this turkey.) It is very simple to analyze; in fact it is the only engine for which we can solve the equations exactly. It provides a benchmark, against which we can measure the performance of real engines, and so understand the importance of slings and hinges.

The second hypothetical engine shown in figure 3.4 has a hinged counterweight, but no sling. This engine (case II) performs better, but is still not very good (a lemon, instead of a turkey). Analyzing case II shows us the benefit of a hinged counterweight. Then we move on to the mangonel (case III) and the trebuchet (case IV). As always, I will float you over the mathematical jungle, but you will not miss out on the underlying physical principles. El Comprendo is the destination; the journey there will be comfortable and painless.

To be specific, I have chosen the following "default" values for some of the parameters. These are realistic values for large medieval counterpoise siege engines. Counterweight mass is to be M = 10,000 kg (10 tons), and projectile mass is m = 100 kg unless stated otherwise. The length of the main beam is that of the short lever arm plus long lever arm: L + ℓ = 12 m, or 40 feet. The mass of the main beam is M_b = 2,000 kg. This is a realistic value for an oak beam that is 12 m long, and thick enough so that it doesn't bend significantly under the enormous forces generated during launch.

To perform the calculations (which get pretty hairy, especially for the trebuchet) I made a few simplifying assumptions in my technical paper about siege engine dynamics (Denny). Compared with the beam and counterweight, the sling mass is very small, so I have taken it to be zero—similarly with the counterweight hinge. For the internal dynamics part of the projectile trajectory, I assumed no air resistance (as for the bow-and-arrow analysis of chap. 1). I assume that the beam is perfectly rigid and that the sling is not elastic. For all our engines there is a trigger to release the counterweight from its raised firing position. For engines with slings there is a second trigger, to release the projectile from the sling. I assume that both these triggers work perfectly—they release the projectile at the right time, and do not influence the motion in any other manner. (For example, the sling does not interfere with projectile trajectory, postrelease.)[6]

Case I: The Turkey

CALLING ON SIR ISAAC to analyze this machine, we find that he obtains exact expressions (given our assumptions) for the launch speed and efficiency. These expressions depend on the engine parameters (lengths and masses). Knowing the launch speed we can find the projectile range. I perform this analysis in the next section, assuming that the projectile is a stone sphere. From the discussion on external bow-and-arrow dynamics you will not be surprised to learn that the best launch angle is about 45°. With our de-

---

6. This is an idealization that can be made very easily by a physicist like me. It simplifies the equations without compromising results. However, for engineers who want to build a working counterpoise engine, the trigger mechanism(s) are a very important consideration. If you check the Web for details of modern counterpoise engine construction, you will find that much effort has been devoted to triggers.

fault parameter values this launch angle gives us a maximum case I launch speed of a little over 20 m s^{-1}. This maximum occurs when the ratio of long-to short lever arm length is $L/\ell = 4$. If we change the projectile size, the weight of the beam, or the launch position (angle θ_0 of fig. 3.4a), then the ratio changes a little, but it is always within a range of 3 to 5. This magnitude for the optimum lever arm ratio also obtains for the more realistic engines. (Lower ratios mean that there is less of mechanical advantage for the lever, and higher ratios mean the counterweight is not enough to accelerate the projectile very quickly.)

The efficiency ε of this engine is

$$\varepsilon = \frac{mv^2}{2Mg\ell(1+\sin\theta_0)}$$

where v is launch speed and, remember, g is the acceleration due to gravity at the earth's surface. From this equation you can see that efficiency increases for heavier projectiles (this result also holds true for the better engines, though the equations for efficiency change). Plugging in the optimum numbers we find that case I engine efficiency is a pathetic 4%. I guess turkeys are not good fliers. This is why the case I engine never found its way onto medieval battlefields. It serves a purpose here, though, by showing what a hill has to be climbed by counterpoise siege engineers to make efficient machines.

Projectile Range

FOR DENSE SPHERES, such as our stone projectiles, air resistance is well understood. The drag force that acts on the projectile increases as the square of projectile speed through the air, and is proportional to sphere area. The drag coefficient for spheres is well known. So, it is a standard exercise in undergraduate physics to set up and solve the equations of motion. Setting up the equations is not too hard, but solving them exactly is not possible, even in principle. They must be solved numerically, by computer number crunching. When this number crunching is done, we can plot projectile range (X_0) as a function of the launch angle (α). The results are similar to the much simpler case of ballistics in a vacuum. In chapter 1 I wrote down the range equation for this vacuum case: $X_0 = (v^2/g)\sin(2\alpha)$. From this expression we saw that maximum range occurred if the projectile was launched at 45°. Compare this

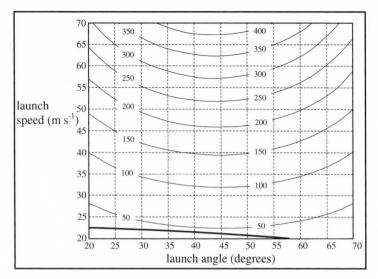

FIG. 3.6. *Range contours (in meters) for a spherical stone projectile. So, for example, a projectile with launch speed of 55 m s⁻¹ (123 mph) launched at an angle of 30° above horizontal will fly a distance of 250 m. Maximum range for a given launch speed occurs when the launch angle is about 45° (slightly less, because of air resistance). The bold line near the bottom shows how our case I engine launch speed depends on angle. This dependence reduces the optimum choice of launch angle to about 40°. Even for this best choice, the maximum range is less than 50 m.*

dependence on launch angle with the number-crunched graph of range versus launch angle, shown in figure 3.6. The range contours are very nearly the same as those for a projectile in a vacuum. This is because air is much less dense than stone, so the projectile is hardly influenced by the atmosphere, unless it is moving very fast.

Also in figure 3.6 you will see a graph of $v(\alpha)$ for the case I engine. I said that this simple lever-arm engine can be solved exactly, but I didn't write down the equation showing how launch speed depends on launch angle, simply because it is quite messy. Instead it is plotted in the figure, which makes the dependence much easier to appreciate. You will see that launch speed is a maximum at 0° launch angle, in other words, if the projectile is launched horizontally. For increasing angles the launch speed falls off slowly. This makes sense physically. The counterweight has more time to accelerate the projectile if the projectile is in contact with the beam (sitting in the spoon at one end) until $\alpha = 0°$. It is released sooner for larger launch angles (stare at fig. 3.4b for a while, and this will become clear to you), and so there is less acceleration and a lower launch speed.

Here is another example of a trade-off. We want a high launch speed, because this gives a longer range (see the range contours of fig. 3.6), but we don't want zero launch angle, since this gives zero range. There is a compromise value, which we can read from figure 3.6. This value—the optimum choice of launch angle to give maximum range for case I—is about 40°. (See how close the curve for $v(\alpha)$ gets to the 50-meter range contour. It is closest at about 40°.) In fact, the turkey flies for only about 45 m (50 yards) which is much less than the ranges reported for real historical siege engines, and the ranges observed for reconstructed modern siege engines. Enemy archers would pick off the crew of a case I engine before they could bring their engine within range.

Figure 3.6 is useful for all our siege engines. For cases II, III, and IV the manner in which launch speed depends on angle is different (the equations are different and much more complicated), but the general trend is similar to that of case I. Range falls off as launch angle increases. Consequently the optimum launch angle remains at about $\alpha = 40°$ for all siege engine types. So, we proceed as follows. Calculate the launch speeds for these other siege engines, then refer back to figure 3.6, and see what range we get with a 40° launch angle. This is the maximum range for the engine.

Case II: The Lemon

FROM HERE ON, the math gets complicated. This is because the number of variables increases. Physicists use the term "degrees of freedom" to describe the complexity of a problem. The equations that describe the internal dynamics of case I engines have 1 degree of freedom: this is the beam angle θ of figure 3.4b. Once the beam angle is specified, then the "state" of the engine is well defined. For case II engines there are two degrees of freedom, as you can see from figure 3.4c: the beam angle θ and the hinge angle ϕ. Both of these need to be known for us to specify the state of a case II engine to see how it moves. Two degrees of freedom means that two equations (coupled differential equations, for those who are interested in such matters) are required to describe the motion.

To establish these equations, we must diplomatically sideline Sir Isaac,

with all sorts of flattery and apologies, smoothing ruffled feathers with sooth-
ing tones, and call in another luminary, from the next century and another
country[7] (Cook; Daintith; Gleick). The Count Joseph Louis Lagrange now
takes center stage, bows, and tells us how to write down the case II equations
of motion. Indeed, I will consult the Count for the remainder of this chapter.
Having obtained the equations, I then must number-crunch them to grind
out the answers. If I then change one of the parameters, say the projectile
mass or the initial beam angle, I must number-crunch them again to get the
new answers. Repeating this process for all kinds of different "initial conditions," I can gain an understanding of what the best parameter values are for
this engine. By "best parameters" I mean the combination of values of initial
beam angle θ_0, long lever arm length L, and hinge length h that produce the
largest launch speed. I am assuming here that the projectile is launched at an
angle of 40°, for reasons already given, and that the counterweight mass, and
so on, are given by the default values. The result of this long process of calculation is that the best case II launch speed is $v = 41.6$ m s^{-1}. Consulting
figure 3.6, you will see that this converts into a maximum range of about 160
m. We obtain this result for an initial beam angle of 55°, long lever arm length
of 7.3 m, and hinge length of 1.4 m.

The improvement in maximum range, over the 45 m of case I, is considerable. We are firing the same weight of projectile, with a beam of 12 m total
length as before, and with the same counterweight and beam weights. So the
huge improvement is due solely to the influence of the hinge, since this is the
only design difference between the two engines.

To see why lemons fly so much better than turkeys, look at figure 3.7. Here
we see the firing of a case II engine, from the instant that the counterweight

---

7. Sir Isaac was easily offended, and very sensitive about the value of his work and
about his role in describing the world mathematically. See the excellent biography by
Gleick (2003). Newton's contemporary Robert Hooke shared these qualities, and added
pugnacity. The two men despised one another, and engaged in a running priority dispute for decades, until Hooke's death. We will encounter Hooke again in the next chapter. Lagrange was an Italian-born French mathematician who reformulated Newton's
Laws in the late eighteenth century. This reformulation is technically equivalent to
Newton's Laws, but it is often more convenient to use and is much more elegant.
Whereas Newton emphasized forces and geometry, Lagrange emphasized energy and
algebra. If Newton represents the seventeenth-century flowering of European scientific inquiry, centered in England, then Lagrange epitomizes the shift of this center to
France in the following century.

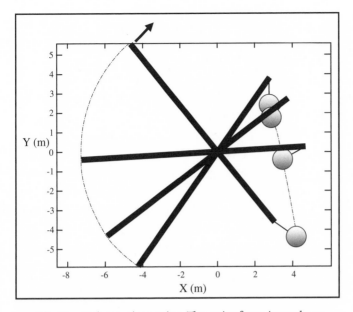

FIG. 3.7. *Case II siege engine motion. The engine frame is not shown. Bold lines represent the beam at four different times: in the starting position, $^1/_2$ second after the counterweight starts to drop, 1 second after, and $1^1/_2$ seconds after. The arrow represents the projectile launch. The initial beam angle is 55° and the launch angle is 40°. Short lines represent the hinge, and circles represent the counterweight.*

trigger is pulled, causing the counterweight to drop from its firing position, to the time (1.5 seconds later) when the projectile is launched from the end of the beam. We see a strobe picture of four superimposed images, separated in time by equal amounts (in this case 0.5 seconds). Note how the beam rotation accelerates, as the counterweight pulls it around. Note, in particular, how the counterweight swings. Initially it swings to the left (in a clockwise direction, about the hinge pivot), but just before launch it swings rapidly in the other direction. This backswing provides a final kick to the departing projectile, and accounts for some of the extra launch speed, compared with case I. As you might imagine, timing is crucial here. If the hinge were a little longer then the counterweight swing would be slower, and the kick would not happen before the projectile was launched. On the other hand, a shorter hinge would lead to a faster swing, and the projectile would be kicked too soon, before it had reached the optimum launch angle. For different choices of long lever arm length, counterweight mass, and so on, we would find that the best choice of hinge length would change from 1.4 m, to provide the kick at the best time. Because the timing is crucial, we find that the launch speed depends very sensitively on hinge length.

The second reason for significant improvement in launch speed can also be seen in figure 3.7. Note how the counterweight drops. Now imagine that it is attached to the end of the beam, instead of hanging from a hinge. You can see that the counterweight motion would have a much greater horizontal component. This horizontal component of counterweight motion is wasted energy. The projectile is supplied energy from the beam, which gets it from the counterweight, which is given stored gravitational energy by the engine crew when they raise the counterweight to its firing position. As the counterweight drops, the stored energy is transferred to the beam. However, if there is no hinge then some energy is converted into counterweight horizontal motion, instead of vertical motion, and so is not available to the projectile. This lack of a counterweight hinge is the only reason why our case I hypothetical siege engine is inferior to our case II hypothetical engine. It is also why the mangonel is inferior to the trebuchet, as we will see.

The optimum choice of parameters (55° initial beam angle, and so on) leads to the least possible counterweight horizontal movement, between the starting position and the launch position (subject to the kick being provided at the right time).

Further calculations show that the efficiency of our optimum case II engine is about 14%—three and a half times better than case I, but still nothing to write home about.

Case III: The Mangonel

NOW AT LAST WE come to a real machine. Here we will assume that the mangonel has a slit trench dug beneath it to accommodate the swinging sling, instead of a rail along which the projectile initially slides. We call again on the Count Lagrange to supply us with the engine equations of motion and on the computer to solve these equations. Once more there are two degrees of freedom, as you can see from figures 3.4d and 3.4e. This time the degrees of freedom are the beam angle θ and the sling angle ψ. Again we tinker with the engine parameters (long arm length, L; initial beam angle, θ_0; and sling length,

r). I will spare you the details and provide the summary, before discussing the underlying physical principles.

Unlike case II we do not obtain a single set of engine parameter values that is optimum. If, for example, I choose the sling to be 5 meters long, then, sure enough, there is a best choice for the other two parameters. ($\theta_0 = 20°$ and $L = 8.5$ m gives the fastest launch speed, of 38 m s^{-1}. This converts to a range of 140 m.) However, if I increase the sling length (r) a little, then the launch speed increases (for a different best choice of the other two parameters). If I keep on increasing r then the launch speed keeps increasing, until I reach unrealistically large sling lengths. For example, if $r = 10$ m then launch speed increases to 56 m s^{-1} (280 m range). This is unrealistic, in the absence of slide rails, because the slit trench would have to be very deep to permit the sling to swing freely. For a 5-m sling the trench would have to be 0.6 m deep (2 feet)—eminently manageable—but for a 10-m sling it would have to be 6 m deep.

We see here the limitations of the mangonel without slide rail, but we also see the advantage of mangonel over the case II engine. The mangonel efficiency is 21% for $r = 5$ m, increasing to 32% for $r = 10$ m. So, if we can overcome the sling-length problem, the sky is the limit, as it were. Furthermore, there is no tricky timing problem here, no kick that needs to be provided by a hinged counterweight, to produce high launch speeds. The mangonel motion is smooth, compared with case II. The projectile acceleration is not jerky. This has logistical advantages to which I alluded earlier: smooth action means consistent projectile grouping on the target. Also increased efficiency means less postrelease rocking, and so quicker reloading. Yet another advantage over case II is the following one, which we have not met before.

Suppose you are the Lord Rupert Headoverheels, a fourteenth-century chivalrous knight, and you have been smitten by the lovely Fiona McGonegal, of Stirling, Scotland. She has been kidnapped by a nefarious ne'er-do-well of evil intent (Baron Frankenstein, say) and you are besieging his fortress. You bring down a tower in his outermost curtain walls with your mangonel (no hinged counterweight—sorry Fiona) and send in your men-at-arms with scaling ladders. A crisis suddenly arises that obliges you to send mangonel projectiles to the inner wall, further away. You can increase range by reducing projectile weight, as we have seen, but that is not enough. So you adjust the sling length and increase range that way. Then you are called back to breach the outer walls. This requires much heavier projectiles than you used when bringing down the tower, because the walls are very strong. So you adjust sling length and increase the counterweight to compensate. Wall breached, men-at-arms pour in, Baron vanquished. Fiona elopes with Clayton, the mangonel chief engineer. C'est la vie.

But you get the point: a sling provides the siege engine operator with an extra option, should the mangonel engineers be asked to change tasks. Projectile range can be quickly adjusted, or projectile weight changed for the same range. It would be possible, I suppose, to make these changes when using engines without slings, but it would be much more difficult.

To see why case III engines perform so much better than case I engines—why adding a sling increases range—consider figure 3.8. As in the last figure, we have a "strobe" effect of four superimposed images, spaced equally in time throughout the internal stage of the projectile trajectory. In this case the time from start to launch is 1.3 seconds, the sling length is 5 m, and the initial

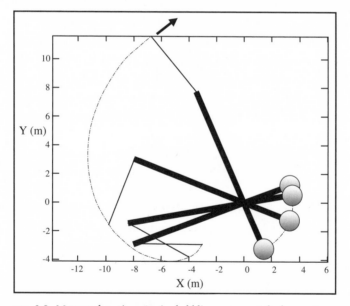

FIG. 3.8. *Mangonel motion. Again, bold lines represent the beam at four different times: initially in the starting position, finally in the launch position (with the projectile direction shown by the arrow), and at two times in between. The initial beam angle is 20° and the launch angle is 40°. Thin lines represent the sling, and circles represent the counterweight.*

beam angle and lever arm length are about optimum for a 5-meter sling. Note that, when the projectile is launched, the beam and sling almost line up. It is as if the beam and sling together form a longer lever arm, with consequently increased lever mechanical advantage. Yet they lack the disadvantage of such a long lever arm (increased beam weight). In fact, further calculations show that this simple explanation does indeed account for most of the increased launch speed found in the mangonel, compared with the case I engine. Compared with our case II machine the mangonel is tactically more flexible, with a smoother action and greater rate of fire.

Case IV: The Trebuchet

HERE THE ADVANTAGES of both hinged counterweight and projectile sling are added to the basic case I engine. The result, as we will see, is stunning (literally, perhaps, if you happen to be on the receiving end). The analysis is also much more complicated because the trebuchet has three degrees of freedom, and because we want a long sling. This requirement for a long sling means we must include a slide rail along which the projectile initially travels—no sweat for an engineer, but a mathematical headache.

We now have three degrees of freedom because we have put together all the component pieces. Thus the degrees of freedom are beam angle, hinge angle, and sling angle. Count Lagrange supplies us with three (coupled differential) equations of motion, and my creaking and groaning desktop computer cranks out the solutions to these equations.[8] When it comes to varying the engine parameters to determine the optimum combination, we again find that things are complicated, because there are four such parameters: initial beam angle, long lever arm length, sling length, and hinge length.

The complications lie in the mathematics, not the physics, and so I have no compunction about skipping it all, as before, to float you to El Comprendo—a physical understanding of the machine—without pain. It turns out that performance improves as hinge length increases, and so I will assume that the maximum possible hinge length is adopted. Such a hinge will cause the counterweight to just skim over the slide rail without touching it,

---

8. I will not describe to you the complications caused by the slide rail. Simply note, for those of you who happen to be physicists, that the constraint is imposed via a Lagrange multiplier.

as the counterweight passes through its lowest point. Now I choose an initial beam angle—let us say $\theta_0 = 45°$—and then tinker with the remaining two parameters until the computed launch speed is maximized. Later on I will explain how to choose the initial beam angle.

For $\theta_0 = 45°$ it turns out that the optimum lever arm length is $L = 8.7$ m and optimum sling length is $r = 7.9$ m. The maximum hinge length is then $h = 2.4$ m. This produces a launch speed of 63 m s^{-1}, which figure 3.6 tells us converts into an impressive 360-meter range (for the default projectile mass of 100 kg, etc.). The efficiency of this engine is a very respectable 58%. So more than half the stored energy of a 12-ton machine is concentrated into a projectile with less than 1% of this machine weight. You can see the "strobe" diagram for a trebuchet with these optimum parameter values in figure 3.9. Note that the counterweight drops nearly vertically—more so than for the case II engine. Why should this be? After all, in both cases we varied the en-

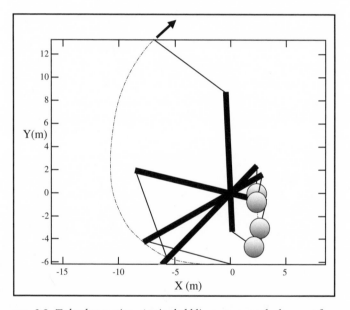

FIG. 3.9. *Trebuchet motion. Again, bold lines represent the beam at four different times: initially in the starting position, finally in the launch position (with the projectile direction shown by the arrow), and at two times in between. The initial beam angle is 45° and the launch angle is 40°. Thin lines represent the sling at one end of the beam, and the hinge at the other end. Circles represent the counterweight.*

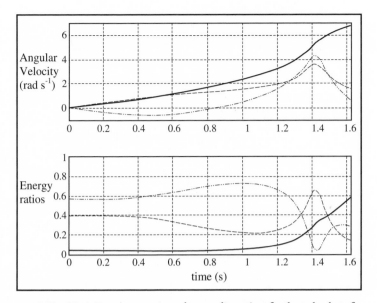

FIG. 3.10. (Top) *Rotation rate (angular speed) vs. time for the trebuchet of figure 3.9. The angular speeds are for the hinge (dash dot), beam (dash), and sling (solid line). Note that both beam and hinge (counterweight) slow down markedly just prior to projectile release, whereas the sling rotation increases.* (Bottom) *Normalized kinetic energy vs. time: counterweight (dash dot), beam (dash), and projectile (solid line). As the motion unfolds, kinetic energy transfers from counterweight to beam to projectile.*

gine parameters to maximize launch speed. The reason for this differing behavior is that, for case II, there is a trade-off between vertical drop (least wasted energy moving the counterweight sideways) and counterweight kick (boosting the projectile at just the right time). Here, in case IV, there is no kick; the action is much smoother, and no compromise is required. So the optimum parameter choice results in a near-vertical counterweight drop with consequent high efficiency.

Also from figure 3.9 you can see how the sling accelerates during the last phase of motion (the internal trajectory takes 1.63 s in this case). The accelerations are plotted in figure 3.10. This figure is very instructive. We see how the beam rotation and hinge rotation are brought nearly to a halt at the moment of projectile release. Thus (as with the torsion spring bow of chapter 1) most of the kinetic energy of the system is transferred to the projectile. This transfer is confirmed by noting the significant rise in sling rotation rate just prior to launch. In the lower plot of figure 3.10 you see how the energy is transferred between engine components. The fraction of total kinetic energy (due to movement of the components) taken up by the beam, counterweight, and projectile are shown. At the beginning, indeed for the first three-quarters of the internal trajectory duration, most of this energy resides within the counterweight. It is then transferred briefly to the beam, and ends up with the projectile at launch. In other words, as the motion unfolds, energy moves from counterweight to beam to projectile. This energy transfer from one end of the machine to the other is reminiscent of a whiplash, where energy moves along a flexible body from the heavy (handle) end to the light end, increasing the body speed as it proceeds. Here, though, the continuously flexible whip is replaced by three rigid beams and rods (due to tension we may regard the sling as rigid in this context).[9]

We now have a reasonably complete picture of trebuchet action. It is an efficient machine because it does not divert much energy uselessly into horizontal motion of the counterweight and because, like the mangonel, it has a long sling that increases the lever mechanical advantage. The initial energy put into the engine by human muscle power is stored in the raised counterweight; when the counterweight is released this energy is mostly transferred through the beam to the projectile.

I said earlier that the initial beam angle was chosen to be 45°. How do I know that this choice of angle is optimal? In fact, it isn't. If our trebuchet engineers put more effort into raising the counterweight further, so that the initial beam angle is 55°, then we obtain a higher launch speed and a more efficient machine. For such a configuration, the launch speed is maximized for a somewhat shorter lever arm ($L = 8.5$ m), a marginally longer sling ($r = 8$ m), and a significantly longer hinge ($h = 3.0$ m). Launch speed becomes 67 m s^{-1}, which converts to 380-m range. Apparently, however, such large initial beam angles were not adopted historically. The illustration of figure 3.5 and many other medieval illustrations of large trebuchets (and modern constructions) favor much lower beam angles. Why should this be, if the engine is more efficient at larger angles? It could be that range is not the key factor,

---

9. Chevedden traces the word trebuchet to Medieval Latin *tribracho,* meaning a device with three arms. On the other hand it has also been identified with the Old French *trebuchier,* meaning to overthrow.

as long as it exceeds that of an arrow. The reason may also be connected with stability of the engine frame and with construction cost. For larger initial beam angles the frame is necessarily larger, and therefore heavier and more difficult to construct and transport. Also the trebuchet recoil, which I will soon discuss, would be worse. Perhaps more significantly, the trebuchet rate of fire is quicker if the initial beam angle is lower (less work in raising the counterweight). Altogether these factors provide pragmatic logistical and tactical reasons for limiting trebuchet beam angle. There is more to effectiveness than efficiency.

If the initial beam angle is reduced (from the 45° value of figures 3.9 and 3.10) to 30° then we obtain a launch speed of 48 m s^{-1} (220-m range) for the optimum choice of (L, r, h) = (8.9, 6.8, 1.0) meters. The ranges that we have obtained here for trebuchets, whatever the initial beam angle, fall within the limits of claims made for historical trebuchet ranges, and for ranges found by modern counterpoise engines. For a given modern trebuchet I can substitute the engine parameters (beam length, launch angle, projectile mass, and so on) and solve the equations to obtain ranges that agree tolerably well with the ranges actually obtained by this machine. Thus we can be confident that the analysis has captured all the significant features of real trebuchets.

Trebuchet Recoil

AFTER THE PROJECTILE has been released, the trebuchet beam and counterweight continue to move. This movement is erratic until dissipative forces bring these engine components to rest. I have ignored friction in the launch phase of engine motion, because this phase lasts for less than a few seconds. The recoil motion, however, can persist for tens of seconds, and consequently friction plays an important role. So here, in this section, friction is put into the equations. By now you will appreciate that it was tactically important for siege engineers to minimize the length of time that the engine components rock back and forth, because this rocking time contributes to limiting the rate of fire. By including friction in the equations and then examining the resulting motion, we can see what can be done to minimize the effects of recoil.

Count Lagrange informs us how friction can be included in the three equations that govern trebuchet motion. (Sir Isaac interrupts, perhaps with irritation, to point out that his formulation of the laws of motion also permits dissipative forces). My overheating desktop computer can solve these equations, but first we need to know how much friction to put into the system. To establish a realistic level of friction I turned to a Swedish Web site that reports on their experience of constructing and firing large trebuchets (Djurfeldt). Crucially, this Web site includes video footage of a trebuchet being fired and continues showing the motion for several seconds after the projectile has been launched. It is possible to estimate from this footage what values to assign to the friction forces that act at the pivots (the main beam pivot and the pivot between beam and hinge).

Number crunching then tells us that the trebuchet beam and counterweight will continue to rock back and forth for several minutes after the projectile has been launched. The rocking motion can take two different forms. The beam and hinge can form a straight line, so that they rock together, like a pendulum. Or the beam and hinge can rock in opposite directions. This second mode oscillates at higher frequency and with lower amplitude. Here low amplitude is the key, since large-amplitude oscillations may destabilize a siege engine, and may be dangerous for the crew. The mangonel cannot recoil in this second mode because it does not have a hinge, so it must always rock in the higher-amplitude, higher-risk pendulum mode. So, trebuchet recoil provides a "softer landing" than does mangonel recoil. For large engines, this may be an important consideration—another trebuchet advantage.

Recoil considerations may explain an otherwise rather puzzling statement made by medieval writers. They claimed that the best engine performance is obtained if the ratio of long to short lever arm lengths L/ℓ is 5.5 or 6. I have found, for all siege engine types, that a lower ratio is better: in general, between 2 and 4 (as high as 5 for the turkey). I am not alone here; all the reconstructed siege engines have this lower ratio. The explanation of this difference may have to do with efficiency; the higher ratios are more efficient. But there may also be a second reason. The engine recoil amplitudes are significantly lower if the lever arm length ratio is increased, as numerical calculations reveal. The large-ratio engines are not optimal, however; trebuchet range is reduced by about half to 180 m. It may be in certain situations, however, that this loss of range was considered to be a price worth paying.

Leftovers

I WAS ONCE ADVISED by editorial staff at the *European Journal of Physics* (EJP) that an appropriate length for papers published in their journal was about six pages. So why would they choose to publish a 17-page article on a type of machine that became unfashionable 500 years ago? Not because of the historical importance of counterpoise siege engines, that is for sure—physics journals do not, and should not, rate historical significance very highly. The EJP people were interested in this paper because their job is to publish articles that are useful for physics professors, helping them to teach physics to university students. Siege engines provide a convenient and easily visualized vehicle for such pedagogical purposes. A lot of physics lectures can be presented around these engines (and a lot of undergraduate laboratory work, too).[10] Here I have (hopefully) conveyed the essence of these great machines by figuratively taking them apart and showing the basic operating principles, such as energy transfer and lever mechanical advantage, that lead to improved performance.

A couple of odd points remain that I have not discussed, which were of practical importance historically and which I list here for completeness.

• *Wheels.* Sometimes the smaller siege engines were placed on wheels. This was not for transportation, but because it was found, perhaps surprisingly, to significantly improve the maximum range. Why should this be so? In light of the analysis presented here, we might understand such improved range by considering the counterweight trajectory—how it moves from the moment the trigger releases the counterweight, until the projectile is launched. The counterweight horizontal movement will diminish if the siege engine frame is mounted on wheels. This diminution occurs because the frame itself can respond to the forces unleashed during the firing process by moving back and forth. Engine efficiency may not be improved, since energy wasted by counterweight horizontal motion is simply replaced by energy wasted by frame motion. However, the frame will be moving forward at the time of launch, and this adds an extra impetus to the projectile.

---

10. Technically the subject matter includes Newtonian and Lagrangian mechanics, holonomic and nonholonomic constraints, velocity-dependent forces, dimensional analysis, and normal modes. The phenomena displayed include energy transfer, pumped oscillation, double pendulum, and sliding friction. For mechanical engineers, too, there is significant teaching material arising from the study and construction of counterpoise engines: beam theory, materials science, tribology. The medieval engineers, of course, knew none of this, and yet they built practical and efficient machines.

A modern improvement upon this idea is the *floating-arm trebuchet*. The idea of a floating arm arose from twentieth-century enthusiasts (Simulator)—perhaps it is the first innovation in siege engine design for 500 years. With this beast, only the beam pivot is placed upon wheels, rather than placing the whole trebuchet on wheels. The beam can move back and forward along the frame, while the frame remains stationary. This design has the advantage of the wheeled trebuchet, but wastes less energy, since it does not move the heavy frame.

• *Propped counterweight.* Some historical (Chevedden) and modern trebuchets (Armédiéval) were designed so that the counterweight hinge is not initially hanging vertically. Instead the counterweight sticks out horizontally, so providing a greater initial torque, to accelerate the beam more rapidly. The prop mechanism is cleverly designed so that it does not interfere with the subsequent motion of the beam, thus maintaining the smooth action.

REFERENCES

Armédiéval. A large group of French enthusiasts who have constructed many large counterpoise engines. Many pictures of siege engines, including one with a propped counterweight, are available on their extensive Web site at: www .xenophongroup.com/montjoie/armed_en.htm.
Castillo, B. D. del (2003). *The Conquest of New Spain,* trans. J. M. Cohen. New York: Penguin.
Chevedden, P. E. (2000). in *The Invention of the Counterweight Trebuchet: A Study in Cultural Diffusion,* ed. A.-M. Talbot. Dumbarton Oaks Papers, Vol. 54. Available at: http://www.jstor.org/journals/00707546,html.
Chevedden, P. E., L. Eigenbrod, V. Foley, and W. Soedel (1995). *Scientific American* 273:66–71.
Daintith, J., and G. Derek, eds. (1999). *A Dictionary of Scientists.* Oxford: Oxford University Press.
Denny, M. (2005). *European Journal of Physics* 26:561–577.
Djurfeldt, P. http://home.swipnet.se/~w-64205/artillery.html. Patrik has kindly confirmed the conclusions that I have drawn from the recoil observations of his trebuchet.
Gleick, J. (2003). *Isaac Newton.* New York: Random House.
Gravett, C. (1990). *Medieval Siege Warfare.* Oxford: Osprey.
Hadingham, E. (January 2000). *Smithsonian Magazine* 79–87.
Hansen, P. V. (1992). *Acta Archeologica* 63:189–268.
Keen, M. H., ed. (1999). *Medieval Warfare: A History,* p. 115. Oxford: Oxford University Press.

Landels, J. G. (1978). *Engineering in the Ancient World,* chap. 5. London: Constable.

Mapes, G. (1991). *Wall Street Journal* July 30, A1, A10.

Middel. See www.middelaldercentret.dk/us_home.htm. The Medieval Center in Denmark maintains this excellent Web site, which details reconstructions of counterpoise engines, with many pictures, and provides a lively historical text. Figures 3.1 and 3.2 are from this site.

NOVA. Text and pictures are available at www.pbs.org/wgbh/nova/lostempires/trebuchet/builds.html.

Payne-Gallwey, R. (1907). *Projectile Throwing Engines of the Ancients,* pp. 29–30. London: Longman.

Simulator. Go to, for example, www.algobeautytreb.com or www.tasigh.org/ingenium/physics.html for simulators and analyses, including the floating-arm trebuchet.

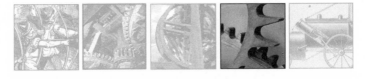

FOUR

PENDULUM

CLOCK ANCHOR

ESCAPEMENT

A Brief History of Timekeeping

MODERN-DAY WORKERS are slaves to time. They rush to catch the 7:42 train or the 6:55 bus to their places of employment, then "clock in" to work and "clock out" when they leave. In the evening they watch one variant or another of the game of football on TV at a preordained time. The game itself depends crucially on timekeeping. The ubiquitous version with a round ball has "time added on." Of the versions played with a prolate-spheroidal ball, the American variety has "time-outs," whereas the rugby league variety has a "final hooter" indicating cessation of play.

It was not always so. Timekeeping to within a second and nanosecond only became essential in the twentieth century. Millennia ago it was important to keep time, but only to the nearest week of the year, to know when crops should be planted, or when the river Nile would flood. Later it became desirable to divide the day or night into equal time intervals of a few hours, for the watch of a soldier on guard duty, or of a sailor at sea. Later still, when industry became large scale and commerce required people to travel and meet each other, it was helpful to know the time to within an hour or quarter hour. The scientific revolution starting in seventeenth-century Europe required ever more accurate timekeeping: measurement of minutes, seconds, and fractions of a second were needed to assist with astronomical observation, maritime navigation, and scientific experiments. This was the period when "national time" was invented, and then international time zones were introduced. And so, we have arrived at the present day. At 10:51 a.m. on the first of June 2005, as I write this line, the computer on which I write it requires timing accuracy of nanoseconds to function properly.

The first timekeeping device (or clock, to estimate time of day, as distinct from calendar, for estimating time of year) was the *gnomon,* a vertical column that would cast a shadow (Bruton, Dresner). Since about 3500 BCE people have estimated time of day from the length of a shadow, be it of Stonehenge or of Cleopatra's Needle (a London landmark, but originally sited at a sun-worshipping temple in Heliopolis, Egypt). An improved version of shadow timekeepers, the sundial, appeared about 1500 BCE. A Roman modification in about 300 BCE, the *hemicycle,* divided the day into 12 equal intervals, and was fairly accurate, though all these devices required adjustment, accounting for location and time of the year.[1] There were two other disadvantages of gnomons: they had to be positioned accurately, north to south (so most of them were not portable), and, to say the least, they didn't work very well at night.

At night, or on cloudy days, the *clepsydra* (water clock [Bruton; James]) has been used since at least 1400 BCE. Water in a vessel would leak out of a small hole in the bottom, and the level of water remaining told time. Originally no allowance was made for the increased flow rate—due to increased pressure—when the vessel was full, but as water clocks evolved over the centuries they became tolerably accurate. From ancient Egypt they spread across the classical Old World[2] and were widely used in Europe during the Dark Ages to call monks to prayer at the appropriate time of day.

---

1. Pliny the Elder, a Roman historian, complained of a sundial in Rome that was inaccurate because its markings had been cut at the latitude of Sicily.

2. The Roman writer Vitruvius—the same fellow who described waterwheels—gave us a description of Roman water clocks.

There were other devices for keeping time prior to the advent of mechanical timepieces. The candle clock in King Alfred's England told the hour well enough for the purposes of religious observance in the ninth century CE. The *sandglass,* or hourglass, is relatively recent, being first recorded in a fourteenth-century Italian fresco. It is a robust and portable timekeeping device that works equally well day or night, independent of latitude, and requires no external power (whether candle wax or water). The hourglass, in fact, used finely powdered eggshells, rather than sand, because this powder ran through the narrow waist more smoothly than did sand. More accurate than the water clock, hourglasses were the industrial timekeepers of medieval Europe.

Some of these early timekeeping devices overlapped in use with mechanical clocks, by as much as four centuries, but eventually they could not compete.

Early Mechanical Clocks

THE FIRST ALL-MECHANICAL clocks (Bruton, Usher) were powered by a weight that was attached to a rope. The other end of the rope was wound around a horizontal axle, so that the axle would rotate due to the torque produced by the weight. This axle was connected, via a train of toothed gears, to a clock face. The first such clocks possessed only an hour hand to display time visually or to cause a bell to be struck hourly.[3] The minute hand did not appear until about 1680 CE, and the second hand a few years later. These two facts alone say a lot about how timekeeping accuracy dramatically improved during the late seventeenth century, and this improved accuracy is the main thrust of this chapter.

In the thirteenth century, however, time was measured only to the hour. A crucial development in the history of mechanical timepieces was the "train of toothed gears." To connect the weight axle with the clock face, the crown wheel and verge escapement (so called because of the shape) were invented. This *crown escapement,* for short, was so basic to mechanical timekeeping that it lasted 500 years, beginning around 1275 CE. An example of crown escapement is shown in figure 4.1, taken from a late seventeenth-century clock. The key function of the connecting gears is to toll out the suspended weight, so

---

3. Our word *clock* derives from the Latin word for a bell.

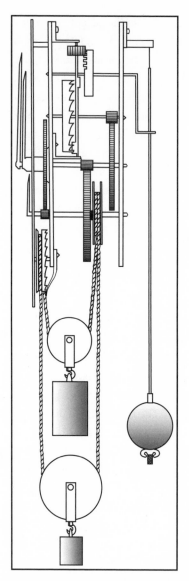

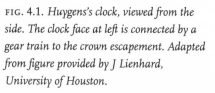

FIG. 4.1. *Huygens's clock, viewed from the side. The clock face at left is connected by a gear train to the crown escapement. Adapted from figure provided by J Lienhard, University of Houston.*

that the hour hand is rotated at a constant, desired rate (an escapement is defined as ". . . a ratchet device that permits motion in steps in one direction only" [Britannica]). I will explain the details of how escapement mechanisms achieve this tolling out later on, when we get to the fourth world-changing machine of this book, the anchor escapement. Mechanical clocks really took off in the later scientific revolution, but the *idea* of a weight-driven clock regulated by an escapement mechanism arose in thirteenth-century Europe.

Pendulum Clocks

AROUND 1582 CE a young medical student by the name of Galileo Galilei observed, in Pisa Cathedral, a lamp swinging from a long chain. He noted, by checking against his pulse, that the period—the time for one complete cycle—of this pendulum appeared to be constant, even when the length of the swing became very small. Although it is possible that Galileo and his son Vincenzio built a pendulum clock (and certainly Galileo recognized the possibility of harnessing the pendulum into a clock), credit for the invention of this machine, successfully combining a pendulum with an escapement to produce a weight-driven timepiece, is conventionally given to the brilliant Dutchman Christiaan Huygens (see box on p. 98).[4] Such clocks were built to his design by Salomon Coster beginning in 1656. (Seven Coster clocks survive today.) These clocks were accurate to within a minute per day and were the first in the world to achieve this milestone. Huygens's later clocks improved in accuracy to 10 seconds per day. Huygens made many significant advances in timekeeping, both theoretical and practical. He invented the balance wheel and spring assembly, used until recently in mechanical watches. He possibly invented the anchor escapement, though Robert Hooke claims to have invented all three. This claim led him into a long and acrimonious dispute with Huygens.[5] For his contributions to timekeeping, if not his conviviality, Hooke also merits a box (see p. 99).

It is possible, even probable, that Hooke *did* invent the anchor escapement (Britannica; Encarta; Landes). Certainly, within weeks of Huygens's patents

---

4. I have chosen my words carefully here. There were many claimants for this key invention. Huygens was the first to patent it, crucially in an age of increasing priority disputes and litigation. This sounds rather modern, doesn't it?

5. Hooke was undoubtedly very talented, and made contributions in many different fields. Unfortunately, he claimed that he originated almost everything he investigated, and this attitude brought him into dispute with many contemporaries. He trod on Newton's toes by claiming to have discovered the gravitational inverse square law. In fact, he may have done so, but not alone, and Newton undoubtedly was the one who derived the consequences mathematically. Arguments over clock mechanisms became widespread and litigious because of prestige and maybe because of the money involved, as we will see.

CHRISTIAAN HUYGENS (1629 – 1695)

Huygens and Hooke would not like to appear in the same box together. I must include them both in this book, however, though this means omitting others who made significant contributions to horology, such as George Graham. But our story here concerns primarily the pendulum clock and anchor escapement, to which these two made key contributions.

Christiaan Huygens was born in The Hague, the son of a well-known Dutch renaissance artist, and was educated at the University of Leiden. The second greatest mathematician, physicist, and astronomer of his age, after Newton, Huygens made major contributions that are still relevant today. His first major work, *Horologium*, described his pendulum clock. This followed from first-principle mathematical calculations concerning the period of a pendulum and how to eliminate circular error, which I discuss in the main text. Disputes with Hooke and others over the spring balance delayed his development of a sea clock. Huygens did not possess Harrison's metallurgical knowledge necessary for this clock to keep accurate time—it varied with temperature. Despite much theoretical and experimental effort over many years, an effective sea clock eluded him.

Huygens improved upon telescope design, and he discovered Titan, the largest of Saturn's moons. He was elected to the Royal Society of London in 1663. Between them, Hooke, Halley, Huygens, and the mathematician/architect Christopher Wren developed the inverse square law of gravitational attraction. Huygens's most important contribution to physics concerned the wave nature of light. This was published as *Treatise on Light* in 1690; it conflicted with Newton's view of light, which Newton thought of as particles rather than waves. Geometrical optics, developed in later centuries, would side with Huygens, and indeed his views on light have been influential in theoretical physics research in the twentieth century. (From the perspective of modern quantum mechanics, both Huygens and Newton were right.) Huygens taught at the University of Paris from 1681 until French intolerance of Protestants caused him to return to his native land. He died in The Hague in 1695.

ROBERT HOOKE (1635 – 1703)

Perhaps the single greatest experimental scientist of the seventeenth century. Over a research career exceeding 40 years, Robert Hooke made significant contributions to a wide variety of fields: physics, astronomy, chemistry, biology, geology, horology and architecture.

There are no universally acknowledged pictures of Robert Hooke that have survived. One modern historian considers that the face shown here is that of Hooke.

The son of a clergyman, young Robert was educated by his father and then from age 13 he attended the prestigious Westminster School, where he shone in many subjects. His education continued at Oxford University as assistant to Robert Boyle, the physicist. Boyle helped Hooke to become Curator of Experiments at the Royal Society in London, from 1662. Four years later Hooke became Chief Surveyor, and helped to rebuild London after the Great Fire. He later became Professor of Geometry at Gresham College. Hooke's name is associated with many inventions: the law of spring elasticity (familiar to every undergraduate physics student), the anchor escapement and watch spring in horology, the universal joint, the photographer's iris diaphragm, and meteorological instruments such as the barometer and anemometer. He wrote *Micrographia*, which detailed his observations of the microscopic world, made with an excellent compound microscope of his own design.

His bad-tempered and lasting disagreements with contemporaries, in particular, Newton and Huygens, arose from his belief that his ideas had been purloined by others who took credit for them. Hooke did not or could not follow up his many intuitive ideas with detailed analyses, and this may have been the cause of his disputes. Sickly as a child, with a crooked, lean frame and unkempt hair Hooke was also plagued by ill health in later life. He died embittered and, like his archrival Newton, intestate.

being granted, the production rights were secured by his splendidly named compatriot Ahasuerus Fromanteel and this patent ushered in the age of English longcase clocks, which dominated horology for a century. William Clement built the first pendulum clock with anchor escapement in London in 1671.[6] Fifty years later George Graham improved the accuracy of pendulum clocks to one second per day.[7] He achieved this by introducing a pendulum that did not change length with temperature, and by inventing the *deadbeat* escapement, a variant (one of many) of the anchor escapement that, despite its name, worked long and successfully. John Harrison, a provincial carpenter and self-taught clockmaker, refined Graham's temperature compensation techniques and added new methods of reducing friction. Most of the clocks that were intended for domestic use were placed in tall (2 meters, to accommodate the pendulum), elegant boxes and so are called *longcase* or *coffin* clocks, later known popularly as *grandfather* clocks. An example of such a clock, with an anchor escapement, is shown in figure 4.2. In figure 4.3 are two modern examples of the closely related Graham escapement.

Note how the improvement in clock accuracy accelerated during this period. From an accuracy of an hour or so in antiquity, to an accuracy of a few minutes (say, one order of magnitude improvement), took a couple of thousand years. The next order of magnitude improvement took a couple of centuries, bringing us to Huygens's era. The next order of magnitude improvement took half a century. This acceleration is due partly to the pendulum (a simple enough device—the clever part was seeing that it could be harnessed to a clock), but mostly it is due to the anchor escapement.

The elegant balance of craftsmanship and technology peaked in the nineteenth century, with beautiful clocks such as the example shown in figure 4.4. (This clock displays technology as art: the gridiron pendulum combines metals with different expansion coefficients in such a way that the pendulum length does not change with temperature.)

Later pendulum clock refinements include a multitude of new escapement devices, a polished pendulum oscillating in a partial vacuum, and a fused-quartz pendulum, the length of which does not vary significantly with tem-

---

6. Hooke denied that Clement was the inventor.

7. Incidentally the division of a day into 24 hours, of the hour into 60 minutes, and the minute into 60 seconds is ancient. It was introduced by the Sumerians, in modern day Iraq, 5 millennia ago.

FIG. 4.2. *(A) An early nineteenth-century English longcase (grandfather) clock, with an 8-day weight-driven movement and a 1-second pendulum. (B) The clock's anchor escapement (the original design invented by Robert Hooke or William Clement). The curved anchor is at the top. Thanks to Renato Zamberlan of Antica Orologeria Zamberlan, Italy, for these images.*

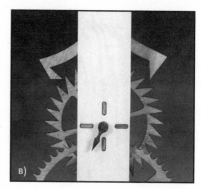

FIG. 4.3. *Modern wooden clocks, with Graham escapements atop each clock. (A) A clock made by Marc Tovar of Utah. (B, C) Images of a clock made by Jeff Schierenbeck of Wisconsin—two American clockmakers who are perpetuating the ancient craft.*

perature. Through these refinements and others, mechanical pendulum clocks remained the most accurate timekeeping machines until about 1930, by which date their timekeeping error rate was down to a few milliseconds per day.

From this brief survey you can see that horology has a rich history, and that from the middle of the seventeenth century there was an explosion of tech-

FIG. 4.4. *A 15-day French mantel clock of the nineteenth century. Note the gridiron pendulum. Thanks to Renato Zamberlan of Antica Orologeria Zamberlan, Italy, for this image.*

nological advances leading to ever more accurate pendulum clocks (and to ever smaller pocket watches). I don't intend to take you through the physics and engineering of all these developments—such an exercise would require several books. Instead I will concentrate on the pendulum clock and, in particular, on the anchor escapement that regulated it, since this device was crucial to future developments. Before taking you along this road, though, I cannot leave the history of clock development without a sidetrack through Longitude Lane, since the subject of longitude estimation was of such importance to the world and since it is so intimately connected with accurate timekeeping.

The Longitude Problem

ON 22 OCTOBER 1707 Admiral Sir Cloudesley Shovel was leading a Royal Navy fleet homeward, in fog. He thought that the fleet was in open ocean, and he ignored those of his crew who suggested that, from the smell of the air, they were nearing land. Four ships of the line foundered on the rocky shores of the Scilly Isles, off the southwest coast of England, with the loss of an estimated 2,000 lives.[8] This disaster was not the first of its kind, and was a consequence of poor navigation, because then there was no method to accurately determine position at sea.

Measuring latitude is relatively easy and has been known to mariners across the world since antiquity. The pole star (Polaris) is pretty much directly over our north pole and so a mariner, in open ocean and out of sight of known landmarks, can estimate his latitude by measuring the angle between the horizon and Polaris. Longitude is a different kettle of fish. There is no star that is the longitude equivalent to Polaris, providing a convenient fixed reference point. The consequences for navigation of this fact were enormous, and not just because of the danger of shipwreck. The European explorers' cartography was plagued by their inability to measure longitude.[9] This problem mattered for reasons of trade with the rapidly expanding colonies in the New World, for example, and with Asia. It also was important in weighty matters of treaty negotiations between empires defining boundaries (Bedini, Pagden).[10] Mariners, navigators, explorers in general *knew* they had a problem and that a solution was possible, but for at least 1,800 years the technology to implement it was lacking.

Seven years after the Scilly Islands incident Parliament passed the Longitude Act and set up a Board of Longitude to look into the problem. A large prize was offered to the person who could provide a satisfactory solution. The Longitude Act (1714) set the prize for longitude estimation depending on per-

---

8. Scilly Isles, silly name, silly man. The admiral, local legend has it, struggled to shore alive but was then killed for the rings on his fingers by a local woman.

9. A Portuguese map of 1541, for example, placed Mexico City 1,500 miles too far to the west. Incidentally the modern techniques of global positioning and mapping, GPS, and remote sensing, are also exquisitely sensitive to timing accuracy.

10. The treaty of Tordesillas in 1494 attempted to specify a line separating Spanish and Portuguese land in the New World. It was ambiguous, and led to confusion and dispute because of the longitude problem.

formance: £20,000 (over $2 million today) for a longitude estimate erring by less than $1/2$ degree, £15,000 for $2/3$ degree error, and £10,000 for 1 degree error. The error was to be that which was incurred upon a sea voyage to the West Indies, 70° longitude west of Greenwich. This was but one of a long list of substantial prizes offered by maritime nations to solve the longitude problem. The first was that of Philip III of Spain in 1598. France and The Netherlands followed suit in the seventeenth century.

Celestial Solutions

HIPPARCHUS, A GREEK philosopher active in the second century BCE, thought of a solution.[11] Imagine Hipparchus is at A (say Athens) and has sent a ship with a competent navigator to B (say Barcelona). They await a lunar eclipse. Hipparchus at A measures the time (he would use a clepsydra) since sunset that the moon first enters the shadow. The navigator at B measures the time since *his* sunset that this same eclipse occurs. The point is that both observers can see the eclipse, and it occurs at the same instant for them both.[12] But sunset does not occur at the same instant; sunset depends on longitude (Barcelona is about $1^1/2$ hours behind Athens). The navigator returns to Athens, and he and Hipparchus compare notes. Let us say that Hipparchus found an interval of $4^1/2$ hours between sunset and eclipse. The navigator reports an interval of only 3 hours. The difference is $1^1/2$ hours. Since the globe turns once every 24 hours, this means that B is one eighteenth of the distance round the world, or 20° west of A.

Note that this method requires two clocks, and that they must be pretty good clocks. The clock inaccuracy—the error in measuring the time interval from sunrise to eclipse—must be small. The estimate of longitude depends on the *difference* between two clock intervals, and so the errors add up. A compound error of 10 minutes in time—not bad for two clepsydras—corre-

---

11. Medieval European astronomers and Arab mathematicians, as well as the classical Greeks, knew the method of determining longitude by timing astronomical events.

12. Because both observers need to see the eclipse, this method will work only for lunar, and not solar, eclipses. In a solar eclipse the moon blocks off sunlight to the earth. Because the moon is so much smaller, it casts only a very small shadow. This small shadow means that only a small area of the earth's surface experiences a solar eclipse. If one happens at Athens, it will not be seen at Barcelona.

sponds to $2^1/2$ degrees error of longitude, or 170 miles (275 km). Also, of course, to use this method Hipparchus needed a lunar eclipse. These were difficult to conjure up at will, and so using the lunar eclipse method required enough knowledge of astronomy to predict eclipses, unless you were prepared to wait around a long time until one happened. Clocks and astronomy were not sufficiently developed in ancient times for this method to be practical. It has been suggested that by 1421 the Chinese made it work for them: they had the best clepsydras in the world and certainly had the astronomical knowledge (Menzies). Even so, the method was inconvenient, because it required the traveler to return home before his longitude (at the time of the eclipse) could be calculated. Put another way, he had to return home to find out where he had been. Also, of course, both locations had to be free of clouds on the night of the eclipse.

A much more serious celestial contender was the *lunar-distance* method, and Johann Müller proposed it as early as 1474. It became one of the two major competitors for the Board of Longitude prize. The basic idea is simple enough. Stars traverse the night sky at the rate of 15° per hour, due to the earth's rotation. The moon traverses the night sky at about $14^1/2$ degrees per hour. This difference occurs because the moon orbits the earth in the same direction as the earth's rotation, so it appears to cross the sky more slowly than the stars. Thus the position of the moon changes against the background of stars. If the star movements and the lunar orbit are known accurately, then the relative orientation of moon and stars fixes an instant of time, just as an eclipse fixes an instant of time. Say the orientation Q happens 1 hour after sunset in Lisbon. So a navigator in some other location measures the interval between local sunset and the time when he sees Q. Knowing when Q happens in Lisbon, he can calculate his longitude relative to Lisbon.[13]

The lunar-distance method is better than Hipparchus's lunar eclipse approach, because the traveler does not need to return to base before he can work out where he has been. However, the method is very difficult in prac-

---

13. Accurate angular measurements in a calm sea (necessary for the lunar-distance method) were made easier by the quadrant, introduced by John Hadley in 1721 (Hooke may have invented it . . .) and later by the sextant. As for how our navigator measures the time interval between sunset and the celestial orientation Q: a conventional pocket watch or small water clock would suffice. The time interval is sufficiently brief (a few hours) that the cumulative errors of these devices—perhaps a few seconds per hour— result in small total navigation error.

tice. First, for six days per month the sun and moon are too close together for lunar-distance measurements to be made reliably. Second, due to the gravitational attraction of the moon and other planets, the earth rotation and lunar orbits are not constant. There are all kinds of perturbations, precessions, and wobbles that need to be measured accurately before this method works at all.[14] These wobble measurements required the accumulated knowledge of decades of astronomical observation. These detailed accurate stellar and lunar observations became available only during the eighteenth century. By then the lunar-distance method was just about workable.[15] Celestial observation methods were workable on land a century earlier (no pitching deck to throw off delicate and meticulous measurements). The well-known French astronomer Cassini had correctly estimated the longitude differences between several European cities via such techniques. But to perform the required measurements, and undertake the lengthy calculations required to make lunar-distance estimates, was beyond the capabilities of navigators at sea. (Imagine trying to use a quadrant to measure the angular position of a star—or even of something more easily seen such as the horizon—when the ship's deck is pitching and rolling. You require measurement accuracy of a fraction of a degree while the deck is pitching back and forth several degrees, and rolling from side to side perhaps 10 or 20 degrees. It just cannot be done.)

The Time Is Right

IN STARK CONTRAST to celestial methods, the other major Board of Longitude competitor—the timekeeping approach—is simplicity itself. The basic idea of using a mechanical clock to determine longitude had been suggested by the Flemish astronomer Gemma Frisius. I can most conveniently illustrate the method with a short story.

You are Admiral Bucephalus Cosmo Diddlysquat, Royal Navy successor to the unfortunate Cloudesley Shovel. You have in common with him two

---

14. Another complication is due to parallax. The moon's position in the sky varies by as much as 2° at any given instant, depending on the observer's location.

15. It is no coincidence that celestial methods and timekeeping methods of longitude determination matured in the same decades, since accurate astronomical measurements require accurate clocks. Indeed, as we shall see, astronomers contributed significantly to horology.

things: his command and a ludicrous appellation. Unlike him you listen to your crew, and are a figment of my imagination. As ABCD you set off from Plymouth dockyard in 1710 to discover exotic islands in the Pacific. To chart these distant lands, you take with you a sea clock, and a book full of Plymouth sunrise and sunset times, for each day of the year. Months later you reach a spot unknown to the world (the indigenous population might disagree), name it after yourself, and head for the nearest Internet café to report back home. To calculate your position, you glance at the sea clock to note the local sunrise time: 6 p.m. (there are no time zones in 1710—your clock has not been reset since you left home port). You check your book to find that sunrise in Plymouth is 7 a.m. that day. The difference is 11 hours; you are $^{11}/_{24}$ of the way around the world—165° east of Plymouth. Simple.

As with the lunar-distance method, though, Beelzebub may be located amongst the minutiae. As Isaac Newton put it, "Once longitude at sea is lost, it cannot be found again by any watch." This economical statement means that the lunar-distance method may be utilized to estimate longitude anew, independent of previous measurements, but a watch or clock, once stopped, would be useless for the remainder of the journey. Worse, a clock running fast or slow, unknown to the navigator, would yield bad longitude estimates. If your sea clock runs fast by 5 seconds per day, and it took you 90 days to reach Diddlysquat Island then your estimate of time difference is out by $7^{1}/_{2}$ minutes. So instead of 11 hours time difference, it is only 10 hours $52^{1}/_{2}$ minutes. This timing error equates to about 1.87 degrees longitude error, which converts to a 125-mile distance error.[16] Another ship setting out to join you at your reported position finds nothing but a bunch of ocean.

But getting a sea clock to be even that accurate was difficult. We have seen that by 1725 the best pendulum clocks could keep time to within 1 second per day. These clocks, however, did not work at sea: the pitching and rolling motion of the ship would upset the regular beats of the pendulum, hopelessly throwing off the timing.[17] Pendulum clocks on land were wonderful time-

---

16. The Longitude Prize called for a six-week journey, from England to the West Indies, for which a clock would have to be accurate to within 3 seconds per day.

17. The previous century saw Christiaan Huygens, motivated by longitude problem prize money, produce a pendulum sea clock that did not work well for this reason. He also developed a spring balance sea clock, and a long and bitter priority dispute with Hooke (who may well have invented clock spring regulators—every physics and engineering student has heard of "Hooke's Law" describing the force of springs) and with

pieces, and would remain so in many homes for another 200 years, but on board an eighteenth-century sailing ship they were all at sea.[18] Another solution was needed.

Over a period of forty years John Harrison, an innovative craftsman of low social standing, produced four clocks (regulated by springs, not pendulums, and dubbed H1 to H4) and submitted them to the Board. These sea clocks were tested; they were found to be excellent by the testers but not by the Board. Obstructions were placed before Harrison, delay followed foot-dragging delay, and there have been dark hints of sabotage. The Board was stacked with astronomers—in particular, Nevil Maskelyne, the Astronomer Royal—who favored a celestial solution and did not want Harrison to succeed. The provincial craftsman became embittered and suspicious.

Outside the Board of Longitude, everybody who encountered Harrison's clocks was impressed. The ships' captains who tested them found that they gave excellent estimates of longitude, and that the clocks were robust enough to withstand arduous and lengthy sea voyages. The great French horologist Pierre Le Roy, clockmaker to the king of France, saw H1 in 1738 and pronounced it "a most ingenious contrivance." His Swiss counterpart and rival Ferdinand Berthoud said the same when he saw H1 in 1763. (Following restoration, H1 is still going today.) Early allies of Harrison included the influential Edmund Halley[19] and the exclusive clockmaker George Graham. These people were impressed by Harrison's clocks and were not put off by his relatively low social standing.

Slowly the Board yielded. Maskelyne admitted that sea clocks may have a role to play during the six days per month when the sun and moon are too

---

others. This clock also proved unsatisfactory, because of the sensitivity of the spring to temperature variations.

18. An aging Newton identified the difficulties to be expected in measuring time accurately with a pendulum clock, even on land. The most significant of these is the variation of gravitational acceleration at the earth's surface at different latitudes. For example, an accurate pendulum clock in London (latitude 52° N) with a 2-second period will, when transported to the equator, lose over two minutes a day. Another source of error is temperature. If the temperature difference between London and the West Indies is 10°C then the London clock will lose about 5 seconds per day in the Caribbean. (Pendulum length increases with temperature.)

19. Halley treated Harrison fairly, despite his close association with, and admiration of, Newton and despite his own interest in celestial science and the lunar-distance method.

close together for lunar-distance measurements to be made reliably. Harrison had been drip-fed small amounts of money by the Board to pay for his work, but prize money did not come his way until 1764, with H4. This pocket-watch-sized sea clock (much smaller than the bulky H1) performed three times better than the Board required to merit the full prize. The error in estimating longitude at the West Indies destination was fewer than 10 miles. Even so, the Board grudgingly awarded him only half the Prize—the other half came seven years later after the intervention of King George III ("By God, Harrison, I will see you righted!").

Harrison's H4 sea clock proved a better solution than the astronomers' lunar-distance approach. Harrison had modified his clocks to make them seaworthy, and reduced their size from a 4-foot box to a pocketwatch. His achievement resulted from sheer mechanical and engineering skill, as much as by horological innovation (and character: describing John Harrison as "determined" is gross understatement, like describing Attila the Hun as "confrontational"). The great sea clocks H1 to H4 were renovated by Gould in the 1930s, and he said of H3, after picking it apart and figuring out the strange mechanisms that made it tick: "No. 3 is not merely complicated, like No. 2, it is abstruse. It embodies several devices which are entirely unique, devices which no clock-maker has ever thought of using, and which Harrison invented as the result of tackling his mechanical problems as an engineer might, and not as a clock-maker would" (Sobel).

Although John Harrison won the Longitude Prize, and sea clocks won the longitude competition, other marine chronometers than H4 solved the everyday problem of finding longitude at sea. The quandary was first tackled successfully by Harrison, but Le Roy solved the broader problem more particularly (Mason). Harrison's solution, being unique and fine tuned, was expensive and difficult to copy. It was much cheaper for a ship to buy charts and tables and use the astronomers' lunar-distance method, cumbersome as it might be, than to fork out the money for a Harrison clock. For this reason, the Longitude Board did not wrap up after Harrison won its Prize, but instead kept going until 1828, after subsidizing the work of many watchmakers and other craftsmen and engineers,[20] and disbursing funds in excess of £100,000.

---

20. For example, the Board of Longitude gave the famous mathematician Leonhard Euler £300 for his lunar table calculations.

In 1763 Le Roy tested his marine chronometer at sea.[21] This was the future. Le Roy's chronometers were designed well, upon lines that were capable of being reproduced inexpensively. Harrison's great sea clocks were the product of a loner's genius; they were one-off designs aimed at winning the Prize. Le Roy represented the future of horology because his scientific approach was more reproducible than the skilled craftsman approach of Harrison. It would still be some time before sea clocks would be made inexpensively (principally by Englishman John Arnold, based in part upon Le Roy's horological innovations), so that every ocean-going vessel could have one and be able to estimate longitude at sea, but the technological problem had been cracked. Harrison and Le Roy rate as Heroes of Technology, and so I provide you with brief outlines of their lives (outside of the longitude problem), conveniently packaged in elegant boxes, like their clocks.

Pendulum Circular Error

NOW IT IS TIME to show you how and why the anchor escapement works so well. Recall that this device replaced the centuries-old crown wheel and verge escapement in the 1760s, and initially was applied to pendulum clocks, increasing their accuracy significantly. These clocks were then modified to function with springs rather than pendulums, so that they would remain accurate at sea.

The old crown escapement mechanism was a coarse device that, in practice, consumed quite a lot of power (with significant friction). When attached to a pendulum clock, such as that of Huygens shown in figure 4.1, it would work well only for large pendulum amplitudes. In other words the swing of the pendulum, which is the angle traversed by the pendulum from one extreme of its path to the other, had to be big. Such a large swing would be inconvenient for longcase clocks, since it would mean that the clock would take up a lot of space, and would lose the elegant slender form of the grandfather clock (fig. 4.2). From the timekeeping point of view, though, there is a much more significant reason to avoid large pendulum amplitudes.

I wrote earlier of Galileo's observation, in Pisa Cathedral, that pendulums

---

21. Le Roy made several technical innovations for his marine chronometer; his earlier association with George Graham influenced some of these ideas.

JOHN HARRISON (1693 – 1776)

In writing biographical sketches of Harrison and Le Roy, and of the longitude problem, in particular, there is a pleasing and appropriate resonance with my earlier sketches of Smeaton and Poncelet. Once again we have the empirical English craftsman and the analytical French scientist both contributing significantly to their field. Their different approaches led to different results, but each earned his place in the historical records.

Harrison was born in a small provincial town in Yorkshire, England, the son of a carpenter. Despite his lowly birth, the teenager John could read and write, and he craved book-learning. He learned woodworking from his father, and began his career as a carpenter. How he became interested in clockmaking is a mystery, but we know that he built his first clock almost entirely of wood, before he was 20 years old. Several other clocks followed, and his local reputation grew. He built two grandfather clocks with his brother James, who was a skilled craftsman. Of the two, John was the innovator, and two early developments proved of lasting importance: the gridiron pendulum and the grasshopper escapement. The pendulum was formed from iron and brass, which react differently to temperature change. This, and the gridiron structure, resulted in a pendulum length that does not change with temperature—essential to accurate timekeeping. Clearly Harrison had taught himself metallurgy at an early age. The grasshopper escapement is typically one-of-a-kind: a unique and utterly brilliant mechanism that operates with minimum friction. The Harrisons tested their gridiron grasshopper clocks by carefully following star movements night after night. Accuracy: 1 second per month.

The pendulum would not do for a sea clock, however, and by the 1720s Harrison had his eye on the longitude prize. He worked on his prize clocks while battling with the Board of Longitude. In later decades he was ably assisted by his son William, who participated in some of the sea trials. After the eventual triumph of H4 (which the explorer Cook took to sea with him, and pronounced to be excellent) Harrison built H5, despite failing eyesight.

Forever associated with sea clocks, John "Longitude" Harrison died on time, 83 years to the day after his birth.

Pierre was born to be a watchmaker. His father Julien was Horologer du Roi, or clockmaker to the king of France, and in time Pierre succeeded him. The son outshone the father in due course. At age 20 Pierre became a master clockmaker, and rose high in the profession as he learned. He introduced many key horological innovations (including the temperature balance compensation and the duplex escapement) which were reported to the French Academy of Sciences between 1742 and 1769.

Le Roy designed watches rather than sea clocks in his quest to solve the longitude problem. A French marquis (de Courtanvaux) had equipped a yacht, at his own expense, especially for the testing of marine watches. Trials were carried out, as in England, under the direction of astronomers, including Cassini. Le Roy's second machine worked very successfully, varying by no more than $7^{1}/_{2}$ seconds in 46 days. A second successful trial resulted in a double prize from the Academy of Sciences, for the watch and for a memoir he wrote on time measurement at sea, in 1770. Ironically much of the benefit of Le Roy's designs went to English watchmakers (in particular, John Arnold) since, as with Harrison, Le Roy's superb timepieces were not easily produced in volume. Arnold achieved this.

Le Roy had long and unpleasant disagreements with his French/Swiss archrival Berthoud; they accused each other of plagiarism. (Arnold and Earnshaw mirrored this behavior in England.) Le Roy spent much of his life and money on horological problems, and won a third prize from the Academy. He made many fine watches and clocks, quite apart from those associated with measuring time at sea.

swing with a constant period independent of the amplitude. This observation is true, but only if the amplitude is small. Consider, for example, a pendulum swinging with amplitude $\theta_0 = 10°$. (This means that the angle between the pendulum at its lowest point, and at its highest point, is 10°.) In time, air resistance will reduce the amplitude, and let us say that after a few minutes it is reduced to 5°. The period remains constant, because these angles are small. The small-angle period is given approximately by

$$T = 2\pi\sqrt{\frac{L}{g}}$$

Here L is the pendulum length. We have already met g, the acceleration due to gravity at the earth's surface. For a one-meter-long pendulum, the period is two seconds, very nearly. An oscillator such as the low-amplitude pendulum, with period that is constant whatever the amplitude, is called *isochronous* (equal time), because the beats are regular. The problem is twofold. First, for large amplitudes (say $\theta_0 = 30°$) the period is not given by this equation; the equation takes a different form that depends on θ_0. This fact is obviously bad news for clockmakers, and Huygens in particular spent a lot of theoretical and practical effort trying to develop a pendulum that avoided such anisochronicity (horologists call it *circular error*, which sounds a little more user friendly than the physicists' term). Second, a clock would need winding up every few minutes if it depended on the amplitude of a simple pendulum, because air resistance would reduce the swing to nothing very quickly. The crown escapement solved the second of these problems, by regulating the pendulum motion in a manner similar to that of the anchor, to be described shortly. The crown escapement needed large amplitudes, however, which meant that the clock beat was not exactly regular, since any slight change in amplitude converted into a change in period. By contrast, the anchor escapement worked for low-amplitude pendulums, because it interfered with the pendulum movement much less and was subject to much less frictional force.

Raise Anchors

AN OLD ILLUSTRATION of the anchor escapement is shown in figure 4.5. This escapement works to regulate the pendulum motion as explained in the figure caption. Energy is fed into the system by a weight that applies gravitational torque to a toothed gear; the escapement mechanism slowly feeds this energy to the pendulum, in small packages, twice per pendulum cycle (once only per cycle, for George Graham's deadbeat escapement). The anchor escapement action has a twofold effect: it keeps the pendulum beating regularly and it increases the interval between clock windings. In the eighteenth

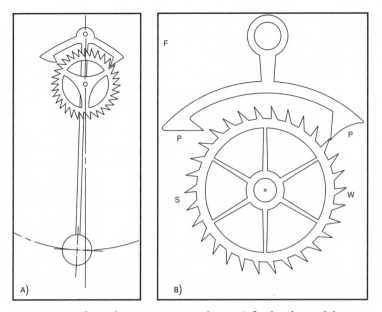

FIG. 4.5. *(A) The anchor escapement (at the top) is fixed to the pendulum, and oscillates with it. The toothed escape wheel is attached via a gear train to a spindle, around which a string is wrapped and attached to a weight (not shown), so that the spindle and the escape wheel will rotate clockwise as the weight descends. This motion is interrupted by the two teeth of the anchor engaging the escape wheel. The anchor shape is such that, when one tooth disengages the escape wheel, the second tooth engages it a very short time later, so that the escape wheel rotates through a small angle (and the clock hand to which it is attached rotates through $^1/60$ of a circle). This action occurs twice per pendulum cycle. The effect of the escape wheel on the anchor gives a small impulse to the pendulum motion, and produces the characteristic tick-tock sound. Note that this diagram is greatly simplified, for clarity. (B) Detail of anchor and escape wheel, from an 1832 encyclopaedia. I thank J. Lienhard for providing these figures.*

century clocks would have to be wound up only once per week, or month (Cook),[22] rather than once every few minutes as for a free pendulum. My tech-

---

22. Graham made two "month clocks" for Halley, which were in use until the beginning of the twentieth century and which ". . . are still keeping time in the Royal Observatory to within a few seconds a week."

nical analysis of pendulum clock escapement movement will show you how this twofold effect arises. Each time the anchor engages the toothed wheel, the mechanism makes a sound: this is the familiar *tick-tock* of a grandfather clock. Literally, the anchor escapement mechanism is what makes a clock tick.

Remember that the anchor menagerie contains many beasts of very different outward appearance. There are dozens of designs for anchor escapement mechanisms. I have already mentioned Graham's deadbeat escapement; there was also the French pinwheel escapement, and Harrison's grasshopper (also called "cricket-head") escapement. Le Roy invented the so-called duplex escapement for watches, which took watch accuracy a big stride forward. There were many others.[23] The precision required of any given anchor escapement design is apparent from the technical drawing shown in figure 4.6.

The astronomer and mathematician George Airy provided an early analysis of escapements (Airy), which he first read to the Cambridge Philosophical Society in 1826. He lacked our modern knowledge of dynamical systems, but his conclusions were valid. As an astronomer, Airy was very interested in timekeeping. The transit of stars across the sky often required accurate clocks for measurement purposes.[24] For example, a star would appear on Tuesday night at the same location as on Monday night, but 3 minutes 56 seconds earlier (Harrison used this fact to test his clocks). Airy's analysis showed that, in general, the pendulum amplitude would vary slightly from beat to beat, and so would the period, depending on the details of escapement design. He showed that it was almost impossible to eliminate both sources of variation (reducing one would cause the other to increase). Overall he favored a type of pinwheel deadbeat escapement.

---

23. Many interesting Web sites contain illustrations of different escapement mechanisms, some of which are animated. I have found that this animation greatly assists in understanding the mechanism. In some cases (such as the grasshopper) this mechanism is quite complicated. For example, see http://mvheadrick.free.fr/escapement .html. Simulations and a more detailed description can be found at www.material worlds.com/sims/PendulumClock/install.html.

24. Airy's interest extended to terrestrial, as well as celestial, timekeeping. He recommended that England should adopt a single time standard. Previously each city and town had its own clock and, since these would not be anywhere near synchronized, you can imagine the havoc of time-tabling transport between towns. This didn't matter so much when roads were bad and transport infrequent and slow, but by Airy's period communications had improved a lot, as the industrial revolution was gearing up.

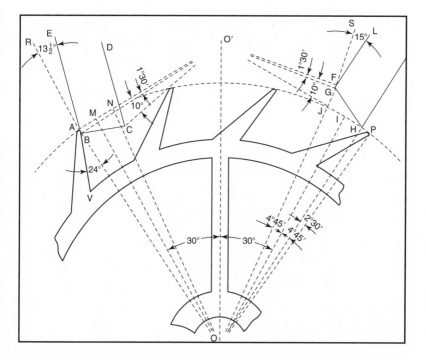

FIG. 4.6. *This drawing gives an idea of the precision engineering required to produce the toothed gears of an anchor escapement mechanism. I am grateful to Christoph Ozdoba for providing me with this image, originally from* Watch Escapements *by James C. Pellaton (1928).*

Dynamical Systems Analysis

TO A MODERN-DAY physicist or applied mathematician, the anchor escapement mechanism is a good example of a nonlinear dynamical system. The nonlinear part makes the mathematics potentially complicated, but also makes the physics interesting. Nonlinearity can lead to chaos—seemingly random and disorganized behavior. Other pendulum applications certainly do become chaotic; the pendulum amplitude and period of such applications are both irregular and nonrepeating. However, I will show you that chaos does not arise for the clock pendulum because of the regulating action of the anchor escapement.

In figure 4.7 I sketch a *phase diagram* for clock pendulum motion. Phase diagrams are important in physics; they permit us to see at a glance important

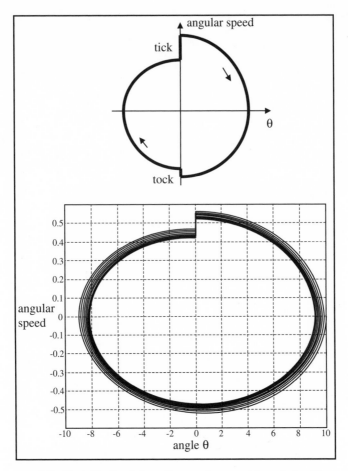

FIG. 4.7. (Top) *Idealized phase diagram for a pendulum clock with recoil escapement. Without friction, the phase plot is an ellipse. Friction turns it into a spiral to the center. The anchor mechanism contributes a small kick twice a cycle, over a short time interval, as the pendulum passes its lowest point. The tick kick increases pendulum angular speed, whereas the tock kick reduces it. If the tick kick is bigger, then enough energy is being input to the pendulum to offset the effects of friction, and a limit cycle results.* (Bottom) *Phase diagram obtained by computer simulation, assuming a deadbeat escapement. In this case the pendulum starts at the right (10° amplitude and zero speed) and spirals inward, only to be restored by the escapement kick. After a few cycles, a stable limit cycle results.*

aspects of the system motion. Our pendulum position is described by the angular displacement θ; that is, how far from the vertical rest position the pendulum is, at time t. The horizontal axis of the phase diagram corresponds to position θ. The vertical axis corresponds to pendulum speed at the same instant. For a simple unregulated pendulum, with no friction to slow it down, the phase diagram is simply a circle or ellipse. This makes sense, if you look at the figure. The pendulum speed is at its maximum when the pendulum is passing its lowest point (θ = 0°), which is what we expect. Also we expect the speed to be zero when the pendulum is furthest away from θ = 0°. Now look at figure 4.7 (top) and the effect of the anchor escapement. When the pendulum passes the lowest point it receives a small kick from the escapement, boosting its speed a little (shown much exaggerated in the figure). This is the clock *tick*. Half a cycle later, when the pendulum is back at the lowest position, it loses a little energy to the mechanism—but not as much as it gained earlier. This is the *tock*. This loss of energy is a practical consequence of many escapement designs, but it is not inevitable. Also in figure 4.7 I show you the result of a computer simulation of clock pendulum motion, but this time assuming Graham's deadbeat escapement. Here there is only a tick, and no tock.

Let me return to the tick-tock (or *recoil*) anchor escapement. Real systems have friction, which means that, without the energy input by the escapement mechanism, the elliptical phase diagram motion would spiral down to the center, corresponding to zero speed with the pendulum just hanging at its lowest position. In the language of dynamical systems, the center of the phase plot is a *stable attractor*. Here "stable" means that, once the pendulum reaches the center, it stays there. Now, with the escapement acting, you see from figure 4.7 that the pendulum does not spiral in to the center, it is given just enough energy each cycle to maintain a stable orbit about the center. This is known as a *limit cycle* (Jordan).

That, then, is the basic reason why clock escapements work. What I will now show you is why the pendulum orbit is stable. After all, injecting energy into other types of pendulum systems can make them go chaotic, which is about as far away from stable as you can imagine. But clock pendulum orbits are stable; this is the beauty of the escapement mechanism. Let me denote the magnitude of the pendulum speed by x_n. This speed is the speed at any chosen point on the phase diagram ellipse. The subscript refers to the tock number: on the 936th tock, the pendulum speed is x_{936}. The escapement influence on pendulum speed can be expressed by the following simple equation (Denny):

$$x_{n+1} = r x_n + k$$

Here r is a number between 0 and 1 and depends on how much friction (air resistance, or friction at the pivot) acts on the pendulum. For no friction, $r = 1$. The number k is the amount of kick that the pendulum gets each cycle. The equation tells us what the pendulum speed will be after the $(n + 1)$th tock when the pendulum is at an angle of, say, 3°, given that the speed was x_n after the nth tock, when the pendulum was at the same 3°. From the equation, you can see how the pendulum speed changes with each cycle. Say that the initial speed is x_0, then from the equation we find:

$$x_1 = r x_0 + k$$
$$x_2 = r x_1 + k = r^2 x_0 + (1+r)k$$
$$x_3 = r x_2 + k = r^3 x_0 + (1+r+r^2)k$$

and so on—you can see that a pattern develops. Given x_0, r, and k, then it turns out that we can work out where the equations lead us. After many ticks (i.e., for large n), we obtain

$$x_n = \frac{k}{1-r}$$

Note that the right side does not depend on n. This means that the pendulum speed becomes constant—independent of tock number—the clock pendulum is orbiting on a limit cycle. The smaller the friction, the bigger the orbit, but even for high friction there is an orbit—the pendulum does not spiral to zero. Now let me suppose that something disturbs the pendulum on the mth tick. Say on the 711th tick our clock pendulum gets a sudden jolt. Perhaps this jolt is due to a maid, Kay, polishing too vigorously a longcase clock box with our pendulum clock ticking away inside. I shall call the jolt K. You can see that, in this case, the system evolves as follows:

$$x_{m+1} = r x_m + k + K$$
$$x_{m+2} = r x_{m+1} + k = r^2 x_m + (1+r)k + rK$$
$$x_{m+3} = r x_{m+2} + k = r^3 x_m + (1+r+r^2)k + r^2 K$$

and so on. Because r is always less than one (if friction is in the system) then for much later times (e.g., many ticks after $m = 711$) the term involving K becomes very small. This means that the pendulum speed x_n is the same as it was without the disturbance K. The pendulum "forgets" it received a jolt: it is stable against sudden disturbances.

There is an important and perhaps surprising lesson here: stability requires friction. Without friction (this is expressed algebraically as $r = 1$), the pendulum speed increases without bound. Without friction, the effect of an occasional jolt is not forgotten; stability is lost. With friction the orbit is stable and the regulated pendulum oscillates with a period given by

$$T = \frac{2\pi}{\sqrt{\dfrac{g}{L} - \dfrac{1}{4}b^2}}$$

Here b measures the amount of friction. Compare this with our earlier equation for the period of an ordinary (unregulated, frictionless) pendulum: you see that the period is longer here. The clock ticks a little bit slower than the rate at which a free pendulum, of the same length, oscillates.

In the next chapter we will see again how important friction is to the stability of a dynamical system. Here we need it for stability, but we do not want too much friction, because this leads to other problems, as seen for the crown escapement. (A pendulum regulated by an anchor escapement does not require large amplitude—3° or 5° is quite sufficient.) Also, large friction means that moving parts wear out quickly. This has other consequences. To reduce friction, some escapement mechanisms require lubrication. But lubricants often gum up in cold weather, and so the degree of lubrication, hence escapement action, depends on temperature. We don't want this; clockmakers want to eliminate temperature dependence, otherwise the unavoidable ambient temperature fluctuations would reduce clock accuracy. If the amount of friction varies with temperature then from the last equation you can see that clock period varies with temperature. Bad.[25]

---

25. Harrison had a characteristically unique and ingenious way around this problem. Some of his escapements required no lubricants at all, and next to no maintenance. They were made of a peculiar dense wood called lignum vitae, which exuded natural oils slowly over time. These oils provided just the right amount of lubrication, and no more.

I haven't written down for you the equation that governs how the regulated pendulum moves. This equation is quite straightforward to derive (if the equations interest you then, see my paper [Denny]) and easy enough to number crunch. In figure 4.7 you see what happens for a deadbeat escapement. The pendulum starts with an amplitude of 10°, at zero speed. (This amplitude is exaggerated, for ease of illustration—real clock pendulum amplitudes are much smaller, as I remarked earlier.) The pendulum motion quickly settles down to a stable limit cycle with amplitude of about 8° or 9°, after a dozen or so cycles. If I had started the pendulum at an angle of 5° it would have spiraled out to the same limit cycle. Note the slight asymmetry in figure 4.7: because of the deadbeat escapement kick at 0° the maximum amplitude is greater on one side (positive θ) than on the other. The pendulum is trying to spiral down to zero, because of the braking influence of friction, but once per cycle it gets kicked back up again.

We can work out how big a kick the escapement gives to the pendulum, in terms of known clock parameters. This turns out to be:

$$k = \sqrt{\frac{2\pi\varepsilon}{\tau}\frac{h}{L}\sqrt{\frac{g}{L}}}.$$

Here τ is the length of time between clock windings (typically one week for a domestic grandfather clock, but one month for Halley's astronomical clocks, and one *year* for later, top-of-the-range clocks), h is the distance dropped by the weight between windings. Recall that the clock is powered by the stored gravitational energy of the weight. Clock efficiency is ε: this efficiency is defined as the fraction of stored energy that is utilized to work the clock—the rest is wasted in noise, heat, frictional loss in the gear train, and so on. For a typical clock, $\varepsilon = 25\%$ is a reasonable value. The equation is interesting, because it tells us that the kick must be bigger for a short pendulum than for a long one. It must be bigger for more efficient clocks than for less efficient clocks, perhaps surprisingly. (This odd relationship may be more readily understood by noting that clock efficiency is not the same as energy input per pendulum cycle.)

For realistic clock parameter values, our equation gives very small kick sizes. These small values show how fine-tuned pendulum clocks must be. Stringent conditions are also placed on the friction force size: it must be very small but not zero.

Winding Down

FOR A PHYSICS TEACHER, anchor escapement pendulum clocks are interesting machines. They are useful vehicles for instructing students about various mathematical methods, and they provide a convenient dynamical system to study, and one that is familiar to most people. Knowing something about the physics and engineering of clock development helps us to appreciate the skill of the many craftsmen during the past five centuries who have contributed to pendulum clock development.

The longitude problem was a bottleneck for human development: it hampered exploration and navigation, and consequently hampered long-distance trade and commerce. The accurate marine chronometers that evolved to solve this problem were the hi-tech marvels of their day. Many hundreds of craftsmen and scientists spent years of their lives seeking this evolution. The pendulum had to be removed, but the anchor escapement remained. That is why I have included the anchor escapement in this book, as one of our most important historical machines. It continued to be used for the most accurate clocks, in one form or another, until electronic timepieces overtook mechanical ones in the 1930s. Anchor escapements lasted a few decades longer in wristwatches (see figure 4.8 for a twentieth-century example), and even today they are still being made, for people who choose to purchase a mechanical timepiece (at considerably more expense than a more accurate quartz watch). Why do they bother? I like to think of it as a tribute to a long tradition of craftsmanship and to human technical ingenuity. A mechanical watch (fig. 4.9) is a thing of beauty.

FIG. 4.8. *A 1950 anchor escapement (from* Watch Escapements *by James C. Pellaton, 1928), designed for a wristwatch. The design differs in detail, though not in principle, from the escapements produced three centuries earlier. I am grateful to Christoph Ozdoba for providing me with this image.*

FIG. 4.9. *A robust, reliable, and very expensive Swiss watch that is much sought after by collectors: a hand-wound Rolex Daytona, with the back removed to show the intricate movement. Thanks to Renato Zamberlan of Antica Orologeria Zamberlan, Italy, for this image.*

REFERENCES

Airy, G. B. (1830). *Transactions of the Cambridge Philosophical Society* 3:105–128.

Bedini, S. A., ed. (1998). *Christopher Columbus and the Age of Exploration*, pp. 436–437, 678–679. New York: Da Capo.

Bowler, P. J. (1992). *The Environmental Sciences*. London: HarperCollins.

Britannica® CD 98 Standard Edition. *Longcase Clock*.

Bruton, E. (2000). *The History of Clocks and Watches*. London: Little, Brown & Company.

Cook, A. (1998). *Edmund Halley*. Oxford: Oxford University Press.

Daintith, J., and D. Gjertsen, eds. (1999). *A Dictionary of Scientists*. Oxford: Oxford University Press.

Denny, M. (2002). *European Journal of Physics* 23:449–458.

Dresner, D., ed. (1998). *The Hutchinson Encyclopedia*. Godalming, United Kingdom: Helicon.

Harrison biographical sketch, sources: Mason *A History of the Sciences;* Daintith, J., and Gjertsen, D., eds. *A Dictionary of Scientists;* and Sobel *Longitude*.

Hooke biographical sketch, sources: Daintith, J., and Gjertsen, D., eds. *A Dictionary of Scientists*. Chapman, A. (2004). *England's Leonardo: Robert Hooke and the Seventeenth-Century Scientific Revolution*. Bristol, United Kingdom: Institute of Physics.

Huygens biographical sketch, sources: Mahoney, M. S. (1980). In *Studies on Christiaan Huygens*, ed. H.J.M. Bos. Lisse, The Netherlands: Swets & Zeitlinger; Daintith, J., and Gjertsen, D., eds. *A Dictionary of Scientists* (see above); Sobel *Longitude* and Usher *A History of Mechanical Inventions* (see below).

James, P., and N. Thorpe (1994). *Ancient Inventions*, pp. 124–126. New York: Ballantine.

Jordan, D. W., and P. Smith (1999). *Nonlinear Ordinary Differential Equations*, 3rd ed. Oxford: Oxford University Press. This book is an excellent introduction to dynamical systems, with many practical examples as well as all the math you need.

Landes, D. S. (1983). *Revolution in Time*. Cambridge, MA: Harvard University Press.

Le Roy biographical sketch, sources: Mason *A History of the Sciences*, Usher *A History of Mechanical Inventions*, and Sobel *Longitude* (see below); horological Web site http://www.datacomm.ch/rbu/.

Mason, S. F. (1962). *A History of the Sciences*, p. 271. New York: Macmillan.

Menzies, G. (2002). *1421—The Year China Discovered the World*. London: Bantam.

Microsoft Encarta(R) Encyclopaedia 2000. CD-ROM. *Clocks and Watches*.

Pagden, A. (2001). *Peoples and Empires*. London: Weidenfeld & Nicolson.

Roy, A. E. (1994). *Orbital Motion*, chap. 2. Bristol, United Kingdom: Institute of Physics Publishing. Measurement of position on the earth and in the heavens is discussed thoroughly in this undergraduate astronomy text.

Sobel, D. (1996). *Longitude*. London: Fourth Estate.

Usher, A. P. (1954). *A History of Mechanical Inventions*, chap. 8. New York: Dover.

Weast, R. C., ed. (1973). *CRC Handbook of Physics and Chemistry*, 53rd ed., p. F-167. Cleveland, OH: CRC Press.

FIVE

CENTRIFUGAL GOVERNOR

Icon

THE CENTRIFUGAL governor (fig. 5.1) is indelibly associated with large industrial machinery, and, in particular, with old-fashioned steam engines. The two flyballs whir around and catch the eye, so that in the confusion of moving parts and metallic noises we see the governor, and pick it out visually from the much larger metallic chaos that surrounds it. Of course the machines to which governors are attached are not chaotic, in part, because of the governor's action. The governor is an automatic feedback device. It was the first such device to be used widely, and it has become an icon of feedback control to many people. It acts as a paradigm in several varied disciplines, including mechanical and electrical engineering, software engineering, neuroscience, and cognitive science. Placing this whirling, almost comical device upon such a pedestal is appropriate, because the

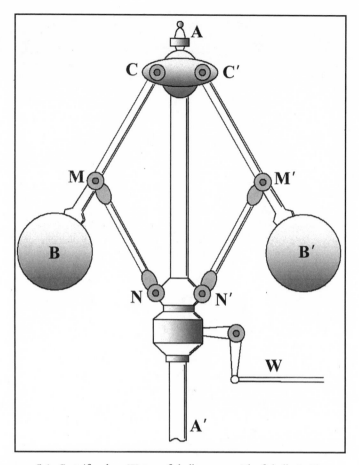

FIG. 5.1. *Centrifugal, or Watt, or flyball governor. The flyballs B, B'
rotate about a spindle AA' that is connected by gears to a steam engine
shaft. Arms MN and M'N' adjust the height of a sleeve, depending on
flyball rotation speed. The sleeve height controls a throttle valve (not
shown) via rod W. The throttle valve in turn controls rotation speed.
I thank Richard Adamek for supplying this image.*

centrifugal governor gave rise to a whole branch of modern engineering science: feedback control.[1]

The reason for the iconic status is clear. Many systems employ *negative feed-*

---

1. The familiar image of a centrifugal governor adorned the front cover of *Scientific
American* (September 1952), for a special edition on automatic control.

back, which is to say a stabilizing influence that dampens out unwanted changes. James Watt wanted his steam engines to run at a constant speed, so he introduced the centrifugal governor as a device to achieve this. The machinery is arranged so that any increase in engine speed causes the governor to shut off steam power to the engine, so reducing speed. Similarly, any reduction in speed causes the governor to react in the opposite direction: it increases steam power and so increases engine speed. Thus changes are stamped out and the engine runs at a constant speed. That is the theory.

This theory can be abstracted easily. Say that a system Q is subject to unwanted changes. A measured change is directed to governor G, which reacts automatically to oppose the change. Now Q might be the temperature of the human body, and G the temperature regulator. You are a jogger and have become hot as a consequence of pursuing this bizarre and unhealthy activity. Your temperature regulator measures the increase in your body temperature and directs the production of sweat, proportional to the temperature increase. The sweat leads to evaporative cooling, which reduces body temperature. So the difference between desired operating temperature and actual temperature, brought on by jogging, has been lessened. Consequently sweat production is reduced or shut off. Normal temperature is regained automatically.[2] Perhaps Q is the speed of a car, and G is the cruise control. You are driving a Ford Thunderbird in cruise control along Route 66 at 75 mph. Your speed decreases as you move from level road to an incline. Cruise control senses this and automatically increases the gas flow to the engine, with the gas flow increase being proportional to the speed difference. Nowadays many of the automatic control devices used in engineering are governed by sophisticated software, instead of a simple mechanical device (for example, fly-by-wire fighter planes). Software allows a much more tailored response to change, and

---

2. The human body also provides us with an example of *positive feedback,* in which a small increase is magnified, instead of reduced. This form of automatic feedback is usually undesirable (e.g., global warming) but sometimes it is helpful. The example I have in mind is childbirth. Our large brains, and hence skulls, make bearing young uncomfortable for humans (to put it mildly), and in historical times rather dangerous. So nature has set up a positive feedback mechanism whereby the increasing pain caused by a baby entering the birth canal gives rise to chemical changes in the mother, causing her automatically to push harder, which increases the pain, and so on. This vicious cycle ends with the birth, and the birth is usually swifter as a result of the positive feedback. We saw a nonbiological example of positive feedback in chapter 3: the arms race between counterpoise siege engine size and castle size.

modern electronic regulators respond to change much more rapidly, but the principle is the same.

The point of these examples is to show you the ubiquity of automatic control, in engineering, biological, and computer sciences. The whole shebang was started off by the centrifugal governor at the end of the eighteenth century.[3] Fifty years later (in the mid-1800s) a crisis developed, a power shortage that might have derailed the industrial revolution. This crisis was averted by scientific investigation of how governors work, mathematically. The importance of the problem is analogous to the longitude problem; it was a technology bottleneck that held us back, economically and in other ways. Resolution of the problem catapulted governors into the engineering limelight and established control theory and cybernetics as disciplines in their own right.

Flour Power

JAMES WATT IS OFTEN given credit for inventing the steam engine and the Watt governor. In fact, he invented neither; his genius lay in improving upon older designs by incorporating his own innovations (separate condenser for the Newcomen steam engine) or by involving someone else's gadget (Thomas Mead's centrifugal governor). Mead was a mechanic and miller, and he is nowadays given the credit for inventing the centrifugal governor.[4] Millers had a problem, and the problem was money. Losing it or gaining it has often been a stimulus to technical innovation, and in this case it was the former.

Windmills have been used to grind flour since the Middle Ages. They work like waterwheels, by grinding the grain between a heavy fixed bed stone and a heavy rotating runner stone. The difference between windmills and waterwheels is the source of power, as we have seen. Wind power is fickle, and is not under the control of the miller. He cannot funnel it down sluices, and adjust the flow rate by raising or lowering gates, as he can with a waterwheel.

---

3. Not quite true. The earliest example of feedback control is probably the water clock float valve, known in classical Greece and Rome. (The modern-day float valve application, perhaps a trifle more prosaic, is in the flush toilet.) The fantail addition to windmills, introduced by the blacksmith Edmund Lee in 1745 to keep windmill sails pointing into the wind, is also an automatic feedback control device.

4. "Mead governor" doesn't quite work for me, however. It sounds like a medieval bartender. I would like to provide you with a biographical sketch for Thomas Mead, but hardly anything is known about him.

Furthermore the change in power that is generated by the wind, and is transmitted by gears to the runner stone, could be rapid, as well as uncontrollable. These power fluctuations at the windmill sails became manifest as changes in runner stone rotation speed. Sometimes the runner would slide slowly over the bed stone, and a few seconds later it might be flying over it. Millers preferred a constant speed, since this gave them consistent flour. But when the speed was too high the separation between runner and bed stone would rise and fall, as the runner bumped along. The resulting flour would be of varying fineness, and such a grainy, inconsistent flour commanded a lower price. So there was powerful financial motivation for millers to regulate runner-stone speed.

The mechanism of centrifugal governors is readily understood at a glance. Two large and heavy balls attached, as in figure 5.1, to connecting rods would fly upward under centrifugal force as axle rotation speed increased. Clearly, flyball angle rises and falls as axle rotation speed rises and falls. Mead's idea (if it was indeed his—he certainly was the first to apply it to windmills) was to use the raised flyballs to press the millstones together. Thus, let us say that a gust of wind causes the windmill vanes to speed up, which in turn causes the runner stone to rotate faster and rise up, and also induces the flyball angle to increase. Now the increased flyball angle causes an increase, via a lever contrivance, of the downward pressure on the runner (or upward pressure on the bed stone), so squeezing them together and countering the bad effect of increased speed. In practical situations such a feedback system worked quite well. Note that this version of flyball control regulates the *output* (flour fineness) but not the power *input*. Mead probably also succeeded in regulating both output and input by using the increased flyball angle differently. Instead of pressing on the stones, the increased flyball angle could be used to *feather* (adjust) the windmill vane angle, via a series of mechanical connections (Bennett). This adjustment trims the runner speed by reducing the windmill vane speed. In other words, it removes the *source* of faster millstone rotation rather than mitigating its *effects*—an altogether better solution.

James Watt knew a good gizmo when he saw one. Within a year of its invention he had, in 1788, adapted Mead's device to the needs of his new steam engine. Harnessed with the throttle valve (which Watt did invent) the new governor worked so spectacularly well that steam engines quickly took over from waterwheels in powering an ever-expanding industrial revolution.

The First Industrial Revolution

THE FIRST INDUSTRIAL revolution was "the most important watershed in the economic history . . . of the world" (Marwick). It occurred in the British Isles during 1750–1850 CE, and was unprecedented and spontaneous. It was also slow, in comparison with the industrialization of Europe, the United States, and Japan, because it was the earliest such revolution, with no precedent to follow. At the beginning of this period, Britain was a country governed by landowners who oversaw a rural agrarian economy; by the end, it was increasingly managed by industrial capitalists who had created an urban manufacturing economy. People were migrating en masse from country to town, and society was losing the agricultural laborer and gaining an urban working class. Factories replaced cottage industries. People were building with iron rather than wood, and their industries were powered by coal rather than water. Cities such as Glasgow and Manchester sprang up in a decade from small towns. The economy, in short, was totally changed, and society followed painfully and reluctantly in its wake. At the end of the industrial revolution, Britain was a radically different country from what it was at the beginning. The same radical changes happened more quickly in other countries that followed suit—in particular, in Western Europe and in the United States—and industrialized during the early nineteenth century.

Henry Maudslay, a manufacturer of wooden pulley blocks, who sold his product to the Admiralty in the ports and dockyards of southern England, summed up the effect and consequences of the industrial revolution in England when he said, in 1802 (Deane): "Driven by a 30 h.p. steam engine the machines made 130,000 blocks a year, cut the labor force from 110 skilled men to 10 unskilled men and saved the Admiralty almost a third of the capital outlay in a year." This quotation encapsulates much of what was happening in those days. Though very unusual a generation earlier, a 30-horsepower engine was common by 1802. The unit *horsepower* had been introduced only a few years earlier by Watt. The increase in usable power brought by the steam engine greatly increased the number and reduced the price of pulley blocks. Finally, production costs were slashed, leading to economic hardship for many workers.

The term *industrial revolution* was first coined (Gardiner) by the French envoy to Berlin, Guillaume, in 1799. It is appropriate that a Frenchman should describe the process as a *revolution*, since France at the time was undergoing a political revolution that followed in part from the American Revolution, and led directly to the Napoleonic Wars. Revolution of the industrial kind is characterized by mechanization of traditional and new industries, and has been defined as a "continuous process of industrial and technological change and therefore [of] sustained economic growth." Why did it happen in politically conservative Britain, rather than in the more forward-looking environments of revolutionary America or France?

The origins of the first industrial revolution have been picked over in minute detail by historians and economists, who see contributory factors in late-eighteenth-century British demography, agriculture, commerce, transport, banking, and innovation.[5] I will concentrate on the last of these, because it is most relevant to the centrifugal governor. Clearly the other factors are important. For example, a country with poor transport infrastructure and large distances between cities (such as the United States during this period) would be at a disadvantage compared with Britain, all else being equal, with its good canals and improving roads, connecting nearby manufacturing centers. In addition, foreign visitors to English factories (such as those of Matthew Boulton and Josiah Wedgwood) always commented on the division of labor, which was much more developed than in their own countries. But innovation is widely recognized as the vital factor. Consider the graph plotted in figure 5.2, which shows the number of patents taken out per decade in England between the years 1630 and 1840 CE. Clearly something happened in the middle of the eighteenth century. Inventions took off. The historians' consensus is that the single most significant invention was James Watt's steam engine.

To see how the steam engine gave rise to a self-sustaining spurt of industrialization, enabling the industrial revolution to kick-start itself, we need look no further than the immediate applications of this engine. The new source of power removed a restriction to growth for existing industries like spinning and weaving, brewing, flour milling, and paper milling. Most importantly, the steam engine pumped water out of coal mines and powered blast furnaces.[6]

---

5. Dr. Johnson, the eighteenth-century lexicographer and famous wit, wrote of the times he lived in: "The age is running mad after innovation."

6. The original application of the earlier Newcomen steam engine was to pump water out of mines. So far as iron production is concerned, the increased blast provided by steam engines permitted furnaces to burn coke instead of the more expensive charcoal, and permitted continuous furnace operation.

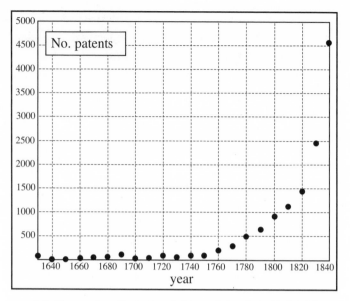

FIG. 5.2. *The number of patents sealed in each decade in England from the 1630s to the 1830s. Data from Deane.*

These applications led directly to cheaper and more plentiful coal and iron, which in turn led to cheaper steam engines with lower fuel costs—another example of positive feedback.

So for our purposes in this chapter the question now is this: what circumstances led to increased innovation in the mid-1700s, and why particularly in England? One explanation is that there was a "fashion for innovation" at that time, epitomized by a small group of open-minded people who were eager to apply the fruits of science to improve the lives of their fellow man (and to get rich doing it).

"Ingenious Philosophers" and Lunatics

THE LUNAR SOCIETY consisted of men (exclusively—they were not so open-minded as to admit women) who were generally settled in their careers, of middle-class background, and respectable but from the dissenting tradition. A "Dissenter" in England then was a person who was not a member of the "established" (State) church—the Anglican Church—but who was instead a Quaker or other nonconformist. Most of the Lunar Society members were nonconformist Englishmen and Scotsmen. At least one prominent member was an agnostic, or perhaps an atheist (discreetly, since such beliefs were not tolerated openly by eighteenth-century society). They were almost to a man politically progressive, which ran against the current of political opinion in England at the time, especially after the French Revolution. The Lunar men (they referred to each other as Lunatics) occasionally included Benjamin Franklin. Whenever he was in England he visited the growing city of Birmingham, in the midlands, where the Lunar Society was based, and where they met regularly during those evenings that had a full moon.[7] Many of these Lunatics are well remembered by scientists and engineers some 200 years after their deaths. The original members of the Lunar Society included Erasmus Darwin (grandfather of Charles), Joseph Priestley, Matthew Boulton, James Watt, and Josiah Wedgwood. There were nine others originally, when the Society first became active around 1775, and more joined over the years.

The Lunatics were interested in science and its application to industry and social life. There was no single field that limited their inquisitiveness: everything from medicine and chemistry to electricity and ballooning interested them (Uglow). Their scientific education had benefited because they had *not* attended the two established universities in England: Cambridge and Oxford. As Dissenters, they were not permitted to enter the hallowed halls of these conservative institutions, which taught theology, Latin and ancient Greek, classics and law, but not much science. Instead, Dissenters attended their own institutions of higher learning in England, or went to Scottish universities, which excelled in science during this period, and which did not require students to uphold the creed of the Church of England.[8]

In general, the Lunar Society men were good friends who supported each other morally and, when required, financially. Wedgwood, the Quaker pottery manufacturer, was always interested in ways to improve and expand his prod-

---

7. It has been suggested that the reason they met at night was to avoid persecution for their dissenting views and liberal politics. In fact, it was simply because they had day jobs. The Lunatics made their way homeward after meetings along lanes with no street lighting, and so needed the moon to guide their way.

8. I should not overemphasize the dissenting component of eighteenth-century British innovators. Many societies other than the Lunar Society met to discuss the latest scientific developments, and many of the Lunar men (Boulton, for example) were not Dissenters. Nevertheless, in England at the time there was a strong correlation, as the statisticians would say, between dissenting views and interest in scientific matters.

ucts. A new method of glazing, or a way to manufacture porcelain, would clearly be of professional interest to him. But the other Lunar Society men joined each other in such pursuits, even when they didn't have a professional stake in the problem at hand. Watt helped Joseph Priestley discover the elemental components of water: hydrogen and oxygen.[9] When Priestley was later hounded by a mob for his dissenting views and support of the French Revolution, he was sheltered by his Lunar Society friends and smuggled out of the country, soon to emigrate to America. Watt, struggling to scrape together funding for his steam-engine investigations, was at his wits' end when his original backer, in Scotland, went bankrupt. In stepped fellow Lunatic Matthew Boulton, and one of the great engineering partnerships of industrial history was formed.

Such were the men of the Lunar Society, who sought to apply science to industry. They supported the American Revolution and were active in the antislavery movement. They were interested in windmills and inoculation, mineralogy and barometers, clockmaking and money, canals and astronomy. James Keir, a Lunatic who made himself rich by manufacturing glass, said after Boulton's death: "Mr B is proof of how much scientific knowledge may be acquired without much regular study, by means of a quick and just apprehension, much practical application, and nice mechanical feelings. He had very correct notions of the several branches of natural philosophy, was master of every metallic art, & possessed all the chemistry that had any relation to the objects of his various manufactures." (Uglow). They all invented and applied. The convivial Ben Franklin, a hero of Boulton's, invented fire grates and water closets, as well as lightning conductors.

Such is the intellectual backdrop to James Watt's world.[10] To see where his centrifugal governor fitted into this, I must tell you a little about the early development of steam engines.

---

9. The first to publish this discovery was Antoine Lavoisier, in France, who heard of Priestley's experiments and rushed to repeat them, and report his results publicly. Lavoisier would later be executed during the Reign of Terror at the beginning of the French Revolution.

10. There is an excellent Web article analyzing James Watt's creativity, and how it arose in part from his upbringing. See www.drl.tcu.edu/Scotland/NorthernLights/watt.html.

Steam Rising

ALMOST NOTHING is known about Thomas Newcomen, except that he developed the low-pressure or atmospheric engine (and learned of the recent scientific advances through correspondence with our old friend Robert Hooke). His engine consisted of a single large piston (between 13 and 75 inches diameter) which cycled through a slow but very long stroke (12 strokes per minute, later increased to 16 or 20, with each stroke typically 6 feet or 8 feet long) (Usher). The effective steam pressure developed was quite low, at about 8 psi. The first engine, as early as 1712, generated $5^1/_2$ horsepower, which is very roughly the same as that of a waterwheel or windmill. The big advantage of Newcomen's engine over these traditional power sources was that it could be placed anywhere. Waterwheels had to be positioned next to a river or stream; windmills needed open country. The first use of steam engines was to pump water out of coal mines, and coal mines were not always conveniently located next to rivers. Later engines would be more powerful and could pump more water from greater depth. So, the Newcomen engine did useful work that permitted mines to be sunk deeper.

These early steam engines were powered by coal fires that boiled water. Steam was let into the cylinder, expelling the air within. A balance beam pulled the piston up to the top of the cylinder. Water was sprayed into the cylinder, causing the steam inside to condense and the cylinder to cool. This cooling reduced the gas pressure inside the cylinder. The piston then was pushed back into the cylinder by atmospheric pressure. Finally, steam was let in once more to begin the cycle again.

Though useful, and superior to the old waterwheel and windmill primary movers, the Newcomen engines were inefficient and burned a lot of coal. This inefficiency did not matter so much in a coal mine, where fuel was plentiful, but it significantly limited the application of steam power. Smeaton, the first "efficiency engineer," whom we met earlier measuring waterwheel efficiencies, applied his methods to Newcomen steam engines. In a series of 130 experiments over three years (1769–1772) he tested engines to see what combination of engine parameters provided the best performance. Smeaton changed piston diameter, stroke rate, boiler size, and so forth, until he came up with a combination of parameters that maximized performance (Mason). This resulted in a doubling of Newcomen engine efficiency. Smeaton developed nothing new; he pursued the same quantitative empirical approach he

JAMES WATT (1736 – 1819)

Born in Greenock, the son of a carpenter, young James was a sickly youth, and for much of his life was ill, or thought he was ill. As a boy he enjoyed mathematics at school, learned carpentry from his father, and, through Watt senior's connections with the Glasgow shipbuilding industry, learned much about navigation instruments, such as quadrants and compasses. James decided to become an instrument maker, and in 1755 was sent to London to learn the trade. This required a long apprenticeship, which he was loath to do, but found an obliging clockmaker who, however, did not pay his unofficial apprentice, but rather was paid 20 guineas by him. John Morgan, the clockmaker, was a master craftsman and a good mathematician, however, and the two got on well with one another. Watt crammed four years of training into one, during which he worked extremely hard and barely ventured outside. In part this was in fear of the press gangs that operated in London at this time, which saw the outbreak of the Seven Years' War with France.

Back in Greenock, safe but sick, Watt set up shop as an instrument maker. Through this business he obtained a commission from Glasgow University to repair a Newcomen steam engine, in 1763. While undertaking this lengthy repair he made his crucial breakthrough in understanding why these old engines were so inefficient. This occurred one Sunday while he was walking through College Green in Glasgow (a plaque marks the spot, for modern-day visitors). In the words of Joseph Black, a chemist at Glasgow University and Watt's friend, the realization ". . . flashed on his mind at once, and filled him with rapture."

Years of frustrating technical and financial problems lay ahead, before his

had used with waterwheels, to get the best out of Newcomen's invention, without contributing anything to the engine development.

Watt's approach was different. His methods were primarily scientific and critical. He would identify a weakness and invent something better, or apply a device not previously used in steam engines. Early in his career he was asked to repair a broken Newcomen engine (Watt, of course, gets his own box in this

continued

dreams of a better engine and financial security were realized. Always suspicious of others' motives, anxious that his ideas might be stolen, Watt said that ". . . the invention has been the product of my own active labor and of God knows how much anguish of mind and body." After his original Scottish backer Roebuck went bankrupt in 1774, Watt took to surveying to look after himself and his young family. Then Matthew Boulton of Birmingham, England, found him. Boulton was diametrically opposed to Watt in outlook and character, but the two got on well for the rest of their productive lives. Boulton was a risk taker, never worrying about debt; he was free spending, outgoing, and convivial. Boulton, the Lunar man whose intellect greatly exceeded that of most industrial manufacturers of the day, backed Watt financially and encouraged and cajoled his brilliant but irresolute partner for many years. Crucially for both their fortunes, he fought for and won a 25-year patent that guaranteed them a virtual monopoly on steam engine production. The first engines went to work in Cornish coal mines, but the partnership was still in debt in 1781. Watt developed the sun-and-planet gears that converted reciprocating piston motion into rotary motion, making his engines suitable for work in factories and not just as water pumps. He adapted the centrifugal governor. By 1800 over 500 Boulton and Watt steam engines had been placed in mills and factories, and Watt could retire a wealthy man.

He was also happier in his later years, despite earlier tragedies that saw the death of his first wife, and of a son and two daughters. He died at 83. We are reminded of James Watt every time we pay our electricity bills, though the kilowatt is not the same as the unit of power he introduced to the world: one horsepower is about 0.746 kW.

book—see above for biographical details) and during the course of the repair he identified the main source of inefficiency. During each cycle of the Newcomen engine the entire cylinder was heated and cooled, which wasted a lot of energy. Watt realized, in one of those flashes of insight that is the stuff of engineering legend, how to remedy this problem—by adding a separate condenser, much smaller than the cylinder, and correspondingly less wasteful of steam energy. This innovation meant that the main cylinder could stay hot throughout a cycle, and so the engine could operate at higher pressure. As a consequence, the Watt steam engine was much more powerful.

Two key machines were added during the many years that it took Watt to realize his version of the steam engine (see box). The throttle valve was a small and simple device that permitted the engineer to control the power of his engine by controlling the amount of steam it received from the boiler. If a heavy load of coal was being lifted out of a mine, the engine power could be increased, and then lowered for a reduced load. This process became automated with the introduction of the second device, our centrifugal governor, which took over control of the throttle valve. It responded automatically to changes in engine wheel rotation speed, and responded faster than an engineer could. The implications of this automatic feedback mechanism for steam engines went far beyond coal mines. A regulated rotation speed meant that steam power could be applied to other uses, which previously were deprived of using it. Thus, for example, the process of spinning yarn to form thread was revolutionized by Watt's engine (the textile industry was one of the most important during this first phase of the industrial revolution). The spinning process required a steady force pulling out the yarn—too slow and the resulting thread is lumpy and uneven, too fast and it breaks.

So the first significant consequence of the centrifugal governor was to assist the spread of steam engines to applications outside the traditional one of pumping water. In the 1780s, steam power rose in importance. Watt's double-acting engine (steam applied to both sides of the piston) with its sun-and-planet gears, throttle valve, and centrifugal governor was the crowning achievement of his career. Steam engine production accelerated. Whereas the older Newcomen engine, in its most efficient form as recommended by Smeaton, consumed 15.87 pounds of coal per horsepower generated, Watt's engine consumed only 6.26 pounds. The consequences of this improved fuel economy were enormous; I will very briefly summarize the most significant of them.

A Watt engine was considerably smaller and required much less fuel than a Newcomen engine of the same power, and this meant more applications. Small factories could find space for an engine. Lots of power could be fitted into the growing factories. This concentration of power meant that factories increase output without requiring correspondingly greater space or a larger workforce. Output increased in many areas of manufacturing. Furthermore,

FIG. 5.3. *(A) An old drawing of one of the earliest steam locomotives, the 1829 "Rocket" of George Stephenson and his son Robert. (B) An English coin celebrating the two-hundredth anniversary of Richard Trevithick's 1804 steam locomotive. (C) An 1875 2-6-0 locomotive of the Virginia and Truckee Railroad, now in a Pennsylvania museum. Thanks to Digital Railroad for this image.*

FIG. 5.4. *A Pennsylvania 4-8-2 "Mountain" locomotive built in 1930. Thanks to Digital Railroad for this image.*

smaller and more fuel-efficient engines meant that they could now be considered for locomotion. All previous engines were fixed at a pit head, or in a factory, but now they could be put on wheels. The age of railroads arrived and railway companies grew fantastically over only a few decades (see figs. 5.3 and 5.4).

Once again a new idea led to positive feedback, exponentially growing the factory output and expanding the industrial revolution into new areas, both figuratively and literally. Thus, the development of steam locomotives led directly to cheaper transport costs for all goods, and, in particular, for iron and coal. Reduced costs of raw materials yielded increased profits and so led to increased growth of the iron and coal industries, which meant that locomotives could be built for less cost, and so there was a railway boom in the second quarter of the nineteenth century. As steam locomotives improved, the rest of the world saw that they were a good idea, and soon railways girded the western world. (By 1840, the United States had more rail track than any other country.)

Equilibrium

SO, IN THE INITIAL stages of the industrial revolution the new steam engines expanded their range of operation because they were small, could be positioned anywhere, and could operate at a steady speed. The Watt governor is responsible for this last characteristic; I will tell you how it works.

The *load* of an engine is the torque that it applies to do useful work—for example, in winding up a bucketful of water from a deep mine. Because of the feedback effect of the governor, an engine will operate preferentially at a certain fixed engine speed. Typically the engine might drive a heavy flywheel, and in this case the engine speed is the flywheel rotation rate. (The flyball speed is proportional to flywheel speed, say it is n times faster, where n is the gear ratio.) The engine speed depends on engine load, in a way that is a function of governor design. For the type of governor shown in figure 5.1, engine speed depends on load as shown in figure 5.5. You will not be surprised to learn that the engine turns more slowly if it carries a bigger load, but the detailed dependence is shaped by governor design. During steady operation, the governor flyballs are raised to a constant angle (the *equilibrium angle*). This angle is large for fast flyball rotation and smaller for slower rotation, due to centrifugal force. Because flyball speed is proportional to engine speed, we therefore see that flyball angle increase with engine speed. The equilibrium angle corresponds to the steady engine speed. In figure 5.6 you can see a curve that is typical of the way that flyball angle increases with engine speed.

Now we have an engine, regulated by a Watt governor, operating at steady speed. Suppose the engine speed changes suddenly, perhaps because the load has changed temporarily (a second bucket of water has been added). To see how the engine responds to this disturbance, under the regulatory action of its governor, we have to set up and solve the equations of motion for this system. This has been done (Denny; Pontryagin), and I will simply show you the results.[11] In figure 5.6 I have assumed that engine speed has been suddenly

---

11. The mathematical description requires some sensible assumptions to be made about how real steam engines operate, but given these assumptions it is not difficult to set up the equations from Newton's Laws. This description was first achieved in the nineteenth century, as we shall see later, and is well known. The first derivation that I saw was in a book by the brilliant Russian pure and applied mathematician Lev Pontryagin. Once the equations are derived, solving them is just math and so, as usual, I omit the details for clarity.

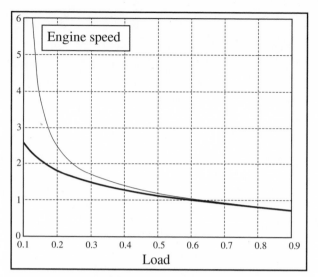

FIG. 5.5. *Engine speed depends on the load torque as indicated (bold line) for the governor of figure 5.1. Different governor geometry can lead to different dependence of engine speed on load (thin line).*

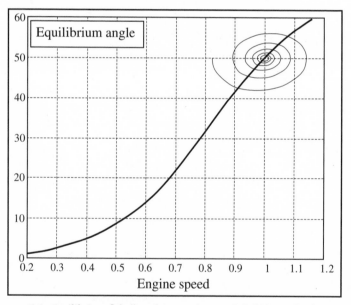

FIG. 5.6. *Equilibrium flyball angle vs. engine speed (bold line), for the flyball of figure 5.1. A different flyball geometry would have a different dependence. If the engine speed is disturbed, it returns to equilibrium as shown (spiral). In this case the engine speed is suddenly reduced from 1.00 to 0.83 (see fig. 5.7). The governor action restores equilibrium after a few seconds.*

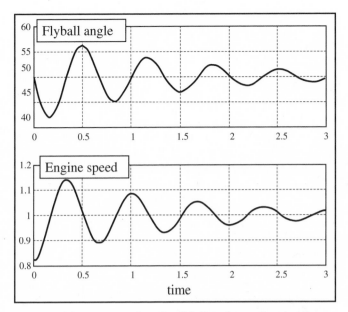

FIG. 5.7. *If the engine speed (and so flyball angle) are disturbed from the equilibrium position, then the governor acts to return it to equilibrium. Here the engine speed has been reduced from a nominal value of 1.00 to 0.83; the equilibrium angle is 50°.*

reduced. The governor responds to this reduction and soon returns the engine to its steady speed, and the flyball angle returns to the equilibrium angle as shown. In figure 5.6 the return to equilibrium appears as a spiral. We can see in this figure how flyball angle and engine speed both change, as the governor acts to restore balance. You can observe the manner in which angle and speed change with time by considering figure 5.7 (which corresponds to the spiral of fig. 5.6). The engine speed increases, then it overshoots the steady value, it slows down again, and once more overshoots, but by a lesser amount. Over a few seconds the engine speed homes in on the steady value. The flyball angle likewise oscillates back and forth, with the oscillations damping down. The governor has done its job; the negative feedback mechanism, stimulated into action by a disturbance from equilibrium, eliminates the disturbance.

This feedback mechanism worked well on the early Watt steam engines, though these machines were crude and unsophisticated compared with their successors. (In fig. 5.8 you can see an old print of a Boulton and Watt engine, with separate condenser, throttle valve, and flyball governor.) The governor

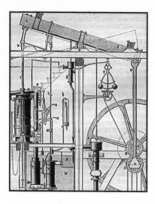

FIG. 5.8. *Watt's 1781 steam engine, showing the key innovations introduced by James Watt. Adapted from an 1878 textbook.*

found a natural home in these engines—it performed better than in its original application of regulating windmills. This better performance is because the steam engine governor required little power, since the throttle valve was small and easy to adjust, so that the governor could change engine speed easily. In a windmill this was not so—changing wind vane speed or millstone separation took a lot more power than changing throttle settings, so the windmill flyballs were big and heavy. The Watt governor flyballs were small and nimble by comparison. But this marriage made in heaven (actually in Birmingham, England) was still in its honeymoon. It would hit a rocky patch later.

Hunting

SEVERAL DECADES later, in fact, and meanwhile there was an enormous increase in the number of Watt steam engines. From the few dozen that Boulton and Watt made for the Cornish coal mines, numbers had increased in England alone to more than 75,000 by 1868.[12] Watt had been dead for half a century, but his brainchild lived on, and was changing the world. I have touched on the expansion of engine applications. In 1868, the engine was an indispensable part of the economy of many countries, in particular, Britain. It powered the nation, and would soon power the world. Engine development continued apace after Watt. Engineering of components improved, so that

---

12. In fact this figure refers to the number of governors, rather than of steam engines.

parts fitted together better, and friction between components was reduced. Engines were made larger and more powerful; the governor size increased also, to better manage the larger throttle valves of the larger engines. The engine-operating speeds increased, and this allowed (indeed, required) a reduction in the size of engine flywheels.

There was a problem. The steam engines were not responding quickly enough to changes in load. Sometimes the response was very slow indeed (the oscillations of figs. 5.6 and 5.7 lasted a long time). Worse than that, sometimes the oscillations did not die down at all, and even increased in amplitude and became irregular. A disturbance in the engine speed would no longer be regulated smoothly. It was as if a puppy on a leash had grown into a powerful dog too strong and willful for his master, who lost control. The governor was no longer governing. This problem, of the engines not responding correctly to a disturbance from equilibrium, was given a name: *hunting*. Perhaps this name originated from the idea of a pack of hounds, when thrown off the scent. Instead of functioning together, and moving in the same direction, the hounds would begin to wander far and wide, trying to pick up the scent once more. This image describes perfectly the seemingly random wanderings of the engine speed (and flyball angle) of a mid-nineteenth-century steam engine that had suffered a disturbance.

The hunting problem was grave, of course, because it threatened the productivity (quantitative and qualitative) of nations. Tens of thousands of engines worldwide were misbehaving, and so the mathematicians, scientists, and engineers of several countries sought a solution to the problem. One promising approach was to design an improved governor that eliminated hunting entirely, instead of simply trying to mitigate the bad behavior, by eliminating the source of hunting. The source, ultimately, was the way in which engine speed depended on load (fig. 5.5). If a steam engine/governor system could be designed so that engine speed was independent of load, then hunting would not arise. Much effort went into designing such engine/governor systems. The German William Siemens and his brother spent over 30 years designing governors, trying to solve the problem of hunting. He described a related problem associated with the Watt type of governor, that of *offset*. He said that the Watt governor ". . . cannot regulate, but only moderates the velocity of the engine, that is, it cannot prevent a permanent change in the velocity of the engine when a permanent change is made to the load upon the engine" (Bennett). An *astatic* governor, one that yielded constant engine

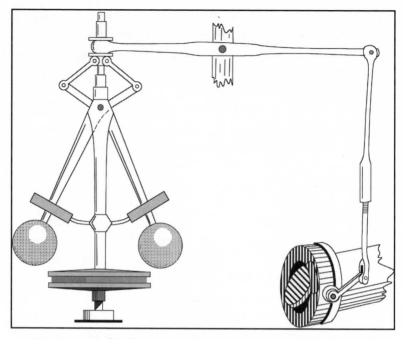

FIG. 5.9. *A crossover flyball governor, said to produce steam engine performance that is approximately astatic. Image from* Discoveries & Inventions of the Nineteenth Century *by R. Routledge, published in 1900.*

speed whatever the load, would solve this problem as well as hunting. The ·French physicist M. L. Foucault[13] devoted much thought to the design of astatic governors. A variant of the Watt flyball governor, which reduces the variation of engine speed with load, is shown in figure 5.9.

The engineers of 1868 were puzzled and worried that the unruliness of their steam engines seemed to be getting worse. Looking back to Watt's time and the intervening decades, they saw that the problems were increasing as engines became better engineered, which seemed counterintuitive to them, and seems counterintuitive to those of us today who are not control engineers. The search for astatic performance failed. Truly astatic engines could not be built, and engines that approximated this state—those with reduced depen-

---

13. Foucault is well known to physicists for developing a very long, low-friction pendulum, which demonstrates the Coriolis force and shows that the earth rotates.

dence of engine speed on load—behaved *worse* than those with a marked dependence, despite the known offset problem. How could the small, crude engines of the earlier days, with a simple governor borrowed from windmills, behave better than the sleek, carefully manufactured behemoths of 1868?

Maxwell and the Rings of Saturn

JAMES CLERK MAXWELL is one of the heroes of physics. Einstein's special theory of relativity arose from his study of Maxwell's equations of electrodynamics and of Newton's Laws. Most people have heard of Newton and Einstein, and know why they are revered. Most physicists put Maxwell on the same pedestal. Here he merits a box, not for his groundbreaking contributions to science but because he was the first person to understand mathematically how Watt governors work (Maxwell 1868). He was asked to look into the problem because of its pressing nature, and because of his known mathematical abilities. It may seem like something of a comedown for the man who analyzed Saturn's rings, and showed why they are stable, to turn his gaze from the heavens to an oily mechanical device that was troubling factory workers. In fact, however, Maxwell was able to use the same mathematical technique in both cases.

The equations that describe the orbits of Saturn's rings are complicated and difficult to solve (without a number-crunching computer). The Watt governor equations are also difficult to solve. So Maxwell simplified the equations by making approximations (I mentioned in chapter 1 that this is an important skill for physicists). In both cases he made the same type of approximation, known as *linearization*. He reduced complicated equations that were impossible to solve exactly to simple equations that were easily solved and that yielded approximate solutions to the original problem. In the case of the Watt governor, these simplified equations told him the answer to the hunting problem.

The mathematical details are not important to us[14]; suffice it to say that

---

14. Maxwell's linearized equation was a third-order ordinary differential equation with constant coefficients, which I here label $c_{0,1,2}$. See the technical references in the bibliography for details. The version I provide here of Maxwell's solution is my own, which is more suitable for general presentation, and is a little more widely applicable than Maxwell's original solution.

JAMES CLERK MAXWELL (1831 – 1879)

Born in Edinburgh, Scotland (a plaque now identifies the house—not a hundred yards from where your author used to live, as it happens), the son of a lawyer, James was a quiet child. This characteristic is often attributed to the fact that his mother died young, of abdominal cancer, when he was only eight years old. James excelled at school, and at Edinburgh University, where he studied Natural Philosophy. During these early years he published research papers on mathematics and physics. In 1850 at Cambridge Maxwell continued gaining praise and prizes for his theoretical studies, and he also branched out into experimental work on color vision.

Best known for his later major contributions to kinetic theory and especially to electromagnetism, Maxwell wrote widely on all kinds of subjects in physics and mathematics. He would become the nineteenth-century scientist with the greatest influence on the twentieth century. In Einstein's words the work of Maxwell was ". . . the most profound and most fruitful that physics has experienced since the time of Newton." After Cambridge, Maxwell was appointed to a chair at Marischal College, Aberdeen, in 1856 and three years later won a prize for his work on Saturn's rings. In his later career he would move on to become professor at King's College, London, and, from 1871, Cambridge. From 1865 to

Maxwell found the conditions that must be satisfied for the governors to successfully stabilize their steam engines. He found three "coefficients," formed from the engine and governor parameters. It is worth writing them down, for reasons that will soon become clear. Do not fret if they appear mysterious to you, because they appeared mysterious to Maxwell's engineering contemporaries as well:

$$c_2 = \frac{b}{m}, \quad c_1 = \frac{g}{L} \frac{\dfrac{d}{L} + \sin^3 \theta_0}{\cos \theta_0 \left(\dfrac{d}{L} + \sin \theta_0 \right)}, \quad c_0 = -2 \frac{g}{L} \frac{G}{I \Omega_0} \frac{h'(\theta_0)}{h(\theta_0)} \sin \theta_0$$

continued

1871 he lived in his family home at Glenlair, near Dumfries, in his native Scotland. It was here that he wrote his definitive paper on governors. This analysis is central to our story, yet forms just a small part of Maxwell's works. In the same year he wrote important contributions to mathematical topology. Maxwell's name is hardly known outside of science, but physicists place him alongside the greatest. The public knows of Einstein and Newton, and immediately identifies these giants with relativity and gravity. Maxwell should be placed alongside them, identified with light. This is because it was Maxwell who unified electricity and magnetism, and showed that they combined to form electromagnetic radiation, of which light is a part.

Maxwell married at age 27 but had no children. He was known for his quirky sense of humor, and his Christian devotion. He died prematurely at 48, the same age as his mother, and of the same cause. The American Nobel Prize–winning physicist Richard Feynman placed Maxwell's contributions in context with the following statement. "From a long view of the history of mankind—seen from, say, ten thousand years from now—there can be little doubt that the most significant event of the nineteenth century will be judged as Maxwell's discovery of the laws of electrodynamics. The American Civil War will pale into provincial insignificance in comparison with this important scientific event of the same decade." This may be hyperbole, or not, but it shows the esteem in which Maxwell is held as a result of the scientific achievements of his short life.

Here θ_0 is the equilibrium flyball angle, L is the flyball arm length (CB in fig. 5.1), $2d$ is the distance CC' in figure 5.1 (so d is the distance from C to the axle), m is the mass of each flyball, and b is the amount of friction in the governor mechanism. These are all governor parameters, along with $h(\theta_0)$, which I will explain shortly. Then there is the load torque G and the engine speed θ_0. The engine flywheel "moment of inertia" is represented by I. This is a measure of the resistance of a flywheel to change speed (try stopping a heavy, rapidly rotating flywheel and you will see what a large moment of inertia it has). The important thing about flywheel moment of inertia is that faster engines in general have smaller flywheels and these flywheels have smaller moments of inertia. So engines that were intended to work at high speeds were designed with low-I flywheels. The function $h(\theta_0)$ describes how the governor flyballs change height as flyball angle θ_0 changes. For the governor of figure 5.1, this function takes a simple form $h(\theta_0) = L \cos(\theta_0)$. This is the form that Maxwell assumed for his analysis. The related function $h'(\theta_0)$ describes how $h(\theta_0)$ changes with θ_0; it is the *derivative* of $h(\theta_0)$.

Well, I did warn you that Maxwell's stability condition may appear a little opaque. The message to take out of the equations is that Maxwell derived three numbers, c_0, c_1, and c_2, that depended, in complicated ways, on governor and engine parameters. He was able to show that the resulting engine/governor system would be stable (no hunting—equilibrium would be restored following a disturbance, as in figs. 5.6 and 5.7) if the following conditions are satisfied:

$$c_{0,1,2} > 0, \quad c_1 c_2 > c_0$$

All the coefficients must be positive: well, the first two certainly are positive, and the third coefficient c_0 is positive if $h'(\theta_0)$ is negative. For the type of governor (fig. 5.1) considered by Maxwell this was the case, and in fact it is true for all centrifugal governors. So, the first condition is satisfied.

The second condition must also hold, and this condition imposes a constraint on engine and governor designs. Violate this constraint and your engine will not be stable—any disturbance will cause it to hunt. Problem solved, at least so far as Maxwell was concerned. Unfortunately, factory owners and engineers were generally not as mathematically astute as Maxwell, and they did not understand his conclusions. For some years after 1868, when the results of Maxwell's investigations were published, hunting continued to plague the steam engine operators of Europe, for this reason. The great man considered that the governor problem was one of dynamical systems theory, rather than a practical economic issue. For nearly a decade this strange situation persisted: the answer was known but not acted upon. The hunting problem continued unabated.

Equilibrium Restored

WHAT THE STEAM engineers needed was a mathematician who could speak their language. They found it, ironically enough, in the form of a Russian engineer whose name is difficult for most non-Russians to pronounce. Ivan

Vyshnegradskii investigated the problem anew (independently of Maxwell). His analysis was more detailed but less rigorous than that of Maxwell, but he reached the same conclusions, and these were published by the French Academy of Sciences in 1876. Expanded versions of his paper were then published in Russian, German, and French (1877–1879) (Vyshnegradskii). The key feature of Vyshnegradskii's analysis is that he was careful to express his conclusions in language that was clear to nonmathematicians. The result was a picture of clarity that many modern-day science writers should strive to emulate.

First, Ivan[15] introduced a parameter v that expresses the steam engine system's *nonuniformity of performance*. This parameter describes how quickly the engine speed changes with load. Mathematically

$$v = \left| \frac{d\Omega_0}{dG} \right|$$

which in English says that the new parameter v is the slope of the curve in figure 5.5. The steeper the slope, the bigger the nonuniformity parameter. Conversely, for the holy-grail astatic system, nonuniformity would be zero, by definition.

Using this new parameter, Ivan was able to express the condition that must be satisfied if we are to avoid hunting behavior in steam engines: $s > 1$ where

$$s = \frac{lbv}{m}$$

The parameter s is referred to as the steam engine system *stability factor*. The condition $s > 1$ turns out to be exactly equivalent to Maxwell's second condition, though it looks much simpler. To explain what this condition means, I can do no better than to paraphrase Ivan's own words:

- Increasing flyball mass is harmful for stability.
- Reducing friction in the governor is harmful for stability.
- Reducing the flywheel moment of inertia is harmful for stability.
- Reducing nonuniformity is harmful for stability.

---

15. Sorry, but I can't keep typing that surname. Nothing against it, you understand, except it takes up so much space.

Recall, if you will, the trends of steam engine development as the nineteenth century wore on: increased flyball mass, better-engineered components and consequently reduced friction, faster engines and therefore smaller flywheel moments, striving toward astatic operation and so reduced nonuniformity. Each and all of these trends led away from stable operations. The puzzling increase in hunting behavior was now as clear as day. Simple changes could be made to turn disgruntled and wayward factory engines into well-behaved workers: don't oil the governor, and replace the flyballs with smaller ones. In the long run better engine designs prevailed, and new governors were developed that took account of Ivan's words.

I have simplified somewhat the monumental and varied work involved in solving the hunting problem, and also the contributions made by Maxwell and Ivan, but you get the gist of it. One great benefit of all this effort, quite apart from happier steam engines and engineers, was the interest generated in applying mathematics to machines. Shortly after the successful explanation of governors, and encouraged by it, mathematicians applied exactly the same type of analysis to turbines, and then to other machines with feedback. The mathematics and the discipline of control engineering were born out of the humble Watt governor.[16]

Nonlinearity

THE HISTORICAL Watt governor story ends with the resolution of the hunting problem. Thereafter steam engine developments continued, other governors were invented to replace the flyball type, and the history of power-production technology moved on. Before leaving this topic, however, I must add parenthetically a final note concerning Ivan's nonuniformity calculation. His analysis, and that of Maxwell, made the simplifying linearization assumption. Nowadays, with our increased knowledge of dynamical systems, we can apply mathematical techniques not known in the nineteenth century. These techniques enable us to estimate, at least approximately, what happens if the linearization assumption is relaxed. In other words, what happens if the flyball angle is a long way from its equilibrium position. To be specific, let me define the actual flyball angle at a given instant of time to be θ, and the differ-

---

16. The modern term *cybernetics*, meaning the study of feedback and control in machines (and animals), was constructed from the Greek word for governor.

ence between this angle and the equilibrium angle to be δ (so that $\delta=\theta-\theta_0$). For the linearization assumption to hold, we must assume that δ is much less than θ_0. It turns out that we can find approximate hunting solutions to the complicated equations of motion even if we permit larger deviations, where δ is not small.

The results are rather complicated, unfortunately, so here I will give you only a flavor, rather than the whole sit-down meal. The stability condition is no longer simply $s > 1$ (i.e., nonuniformity must be greater than one). Instead, the nonuniformity must exceed a function that depends on δ and upon θ_0. For small values of δ this function reduces to 1, so that Ivan's linearization result obtains in this limit, as it should. For larger values of δ the function increases to a value greater than 1, if θ_0 is a moderate value (say 30°). So, for example, the engine/governor would be stable if $s > 1.1$ for $\delta = 15°$, or $s > 1.2$ for $\delta = 20°$. It happens that in this region the hunting solution is not itself stable. It is like the equilibrium of a pin standing on its point—any slight disturbance and it falls over. Here this translates into the flyball motion becoming stable or, just the opposite, becoming irregular, but not forever hunting about an equilibrium angle. On the other hand if θ_0 is large, say 70°, then the stability function behaves differently: it falls below 1 as δ increases. The hunting solution is stable—the flyball oscillates about the equilibrium angle without growing or diminishing. Its hunts forever, never settling down.

This type of varied and complicated behavior is just what was happening to steam engines prior to Maxwell and Vyshnegradskii. After them, steps were taken to ensure that large deviations from equilibrium did not occur, and that any small disturbances that arose were quashed by the governor feedback mechanism. Since then, control theory has not looked back.

REFERENCES

Bennett, S. (1979). *A History of Control Engineering 1800–1930*, chaps. 2 and 3. Stevenage, United Kingdom: Peter Peregrinous.

Deane, P. (1979). *The First Industrial Revolution*. Cambridge: Cambridge University Press.

Denny, M. (2002). *European Journal of Physics* 23:339–351.

Gardiner, J., and N. Wenborn, eds. (1995). *The Companion to British History*. London: Collins and Brown.

Marwick, A., ed. (1980). *The Illustrated Dictionary of British History*. New York: Thames and Hudson.

Mason, S. F. (1962). *A History of the Sciences.* New York: Macmillan.

Maxwell, J. C. (1867–1868). Proceedings of the Royal Society 16:270–283.

Maxwell biographical sketch, sources: Daintith, J., and Gjertsen, D., eds. (1999). *A Dictionary of Scientists.* Oxford: Oxford University Press; O'Connor, J. J., and E. F. Robertson, article on the St. Andrews University Mathematics Department Web site, www-groups.dcs.st-and.ac.uk/~history/Mathematicians/Maxwell.html; James Clerk Maxwell Foundation Web site, http://www.clerkmaxwellfoundation .org/; Wikipedia Web site, http://en.wikipedia.org/wiki/James_Clerk_Maxwell.

Pontryagin, L. S. (1962). *Ordinary Differential Equations.* Reading, MA: Addison-Wesley.

Uglow, J. (2002). *The Lunar Men.* New York: Farrar, Straus and Giroux.

Usher, A. P. (1954). *A History of Mechanical Inventions* New York: Dover.

Vyshnegradskii, I. A. (1877). *Civilingenieur* 28:96–132.

Watt biographical sketch, sources: Deane *The First Industrial Revolution*, Uglow *Lunar Men*, and Mason *A History of the Sciences*. Daintith, J., and Gjertsen, D., eds. (1999). *A Dictionary of Scientists*. Oxford, Oxford University Press.

INVENTIVENESS

Summary of Why and When

I WOULD LIKE TO try to assess the *inventiveness* of the people who developed our five machines, but before doing this it would be useful to place their contributions in context, both technologically and historically. So first I will summarize the whys and whens: *why* I consider each machine to be a world changer and *when* it was most influential.

Bows and arrows were important from prehistoric times for hunting. A bow extends the hunter's lethal range and significantly improves his chances of successfully hunting fleet-footed prey such as a deer or rabbit. A hunter would have found it much easier to get within arrow range of his prey, than to get within spear range. Because of the high-protein content of meat compared with vegetable matter, such as fruit and berries, even small prey species will have greatly enhanced his chances of survival, in particular, in marginal regions of the world. Expanding human numbers led to migrations, and those populations who found themselves on the margins will have survived better with bows. So the northern tundra, the Asian steppes, and the American plains became habitable. I do not know of any quantitative studies that have been performed to support this view, but it seems reasonable to me to assume that a given marginal area would support more humans with bows than humans without bows. So, bows changed human demography since prehistory.

If people are competing for resources then warfare cannot be far behind, and we have seen how, and when, archery has altered warfare and therefore changed the world. Dense formations of infantry or cavalry were vulnerable to archers used en masse: this sounded the death knell for medieval European chivalry from the mid-fourteenth century, as we have seen. Archery spurred the development of long-range siege engines, so the engineers would be able to carry out their work unhindered. Firearms existed side-by-side with bows for a couple of centuries until guns became superior (in range, accuracy, and reliability). This occurred in the sixteenth century in Europe, and somewhat later in other parts of the world. We can date the decline of bows and arrows from this period, since firearms replaced them in both warfare and hunting. So, bows were influential in warfare from antiquity until, say, the sixteenth century.

Apart from changing the world, bows and arrows have given us some more trivial "echoes," as I termed these historical vestiges in the Introduction. One such echo is the two-fingered English gesture, a longbow reference from the Hundred Years' War. Another is the expression "parting shot," yet another is in the surnames Archer, Bowman, and Arrowsmith. Like dinosaur bones, these relics of a once mighty past survive to the present day as intriguing reminders of former glory.

Waterwheels were our prime movers for about two millennia. *Windmills* were influential for about one millennium. The mobility of oxen, and later horses,[1] was needed to plow fields and draw wagons, but where fixed power sources sufficed the waterwheel or windmill was better. These could pump water to drain marshes and mines, and to irrigate fields. They could mill flour and saw wood. The mechanical power sources did not tire, kick, or defecate.

---

1. The horse harnesses of classical antiquity were unsuitable for pulling a plow. The harness rested on the horse's neck, and so traction would block the windpipe. Horses were used as draft animals only after a more suitable harness was introduced later on; this new harness rested on the shoulders.

On the other hand the waterwheel had to be built within a few hundred yards of running water, and the windmill required open land in a windy part of the world. Both machines became efficient and effective. For specific tasks they became essential. Thus, it is hard to imagine how the Dutch polders could have been drained and kept dry without the full-time application of windmills. Equally, only the waterwheel could have powered the first phase of the industrial revolution. Draft animals could play only a minor role in these cases. With the advent of Newcomen's steam engine the task of pumping water out of coal mines was removed from the remit of waterwheel applications. With the introduction of Watt's improved engine the traditional prime movers became largely redundant. So, waterwheels and windmills were important economic factors from antiquity until they were replaced by steam power at the end of the eighteenth century.

Historical echoes include the English name Miller and the German Müller, and I suppose that there was once a *Red Mill* on the Parisian site of the famous *Moulin Rouge* cabaret. However, the most concrete and poignant echoes of windmills and waterwheels are the carefully restored and maintained structures themselves, and the societies that have sprung up in their defense.

Counterpoise siege engines changed the face of medieval Eurasian warfare, with long-lasting consequences for military architecture and organization. Mangonels and especially trebuchets became gigantic. With them, a rock weighing several hundred pounds could be launched accurately over several hundred yards. They would pulverize any defensive structure except the largest and strongest castles. These castles grew large in response. The shape, size, and constitution of castles evolved in parallel with the threat from counterpoise engines. The shear scale of these castles required new levels of logistical and organizational skills. Again, I know of no quantitative study to support this view; such a study might attempt to correlate trebuchet size with, say, the number of stonemasons or architects. However it seems to me to be more than a coincidence that European cathedrals grew large at exactly this time. The same skills needed for large-scale and long-term projects such as castle building could be applied equally well to the majestic steeples and to the airy naves and aisles of gothic and English perpendicular cathedrals.

These large siege engines served only one purpose, and this specificity perhaps explains their relatively brief (three or four centuries) period of dominance. The fully developed counterweight engine appeared around 1200 CE and was eclipsed by weapons using gunpowder in the 1500s.

FIG. 6.1. *Medieval mangonel and trebuchet reconstructed by a Danish museum, with engineers—and others—in medieval costume. Image provided by The Medieval Centre, Nykoebing F, Denmark.*

Counterpoise engines echo today in the reconstructions that festoon French chateaux, Danish museums, English, Scottish, and Swedish fields, and American TV history programs and university physics departments—and in naming the profession of "engineering."

The *anchor escapement,* when attached to a pendulum, gave us accurate clocks on land. The impact on timekeeping followed immediately upon the marriage of these two simple devices. The pendulum had to be thrown overboard, figuratively speaking, when it came to sea clocks, but the anchor escapement remained. The early hand-crafted sea clocks of Harrison and Le Roy, and the later mass-produced marine chronometers, all used this escapement in one form or another. Accurate mapping always has and always will require accurate timekeeping. Accurate navigation by traditional methods also needed accurate timekeeping. The Longitude problem had held back European exploration, colonization, and trade. For better or worse, the anchor escapement played a significant role in expediting these world-changing events. On a more domestic level, grandfather clocks and pocket watches told the time to ordinary people all over the world for several centuries, from the late 1600s until the mid-1900s. The humble anchor escapement was born and evolved in an epoch of rapid change, and lasted 250 years. It still exists, of course, but as nostalgia, as a museum piece,[2] but it ceased to be *important* about 70 years ago.

A multitude of mechanical clocks and watches survive as echoes of the past, but for the one small component that I have written about there is only a single, but resonant, echo. The tick-tock sound of old clocks arises from the action of our anchor escapement. Today it sounds in many a digital clock and watch. Of course, these electronic timepieces need not produce such a sound, but people like to hear a timepiece tick-tock and so some digital timepieces are made to artificially tick-tock. A clock isn't a clock without a tick-tock.

Centrifugal governors regulated the power of James Watt's steam engines. The resulting engine had a dependably constant speed, and this permitted its use in many specific applications (such as spinning yarn) where it previously had not been used. Thus the governor contributed directly to the spread of steam power. Later the engines became unreliable because of the hunting phenomenon. This unwelcome phenomenon arose because of changes in engineering design and because of insufficient understanding of feedback dynamics. The resolution of this problem led directly to the establishment of control engineering as a distinct and valuable discipline. When it comes to estimating the importance of the centrifugal governor it is hard to separate it from the throttle valve and the sun-and-planet gears. Together they enabled the steam engine to power the industrial world, and this is one of the most important technological advances in our history. The centrifugal governor itself flourished for perhaps only 90 or 100 years from the 1780s; thereafter, superior regulators displaced them.

The centrifugal governor echoes today in our word "cybernetics," and as an icon for feedback in many distinct disciplines.

Comparisons

IT SAYS A LOT about our five machines that, though all are now thoroughly obsolete, all are still being made, restored, conserved, and/or operated by en-

---

2. The same might be said of windmills, waterwheels, and the flyball governor. Bows and arrows are still used by hobbyists. Enthusiasts are still making trebuchets. People are fond of good ideas.

FIG. 6.2. *An unusual Belgian medieval archery competition. I thank Olivier Picard for permission to reproduce this image.*

thusiasts, all over the Old and New Worlds. Those unfortunate members of society who are not interested in old machines may consider it quite eccentric to see a bunch of otherwise-normal Danes firing stone projectiles from large wooden engines (in medieval costume—see fig. 6.1). They may wonder what makes an American build a full-scale working replica of a 200-year-old water-powered mill (see fig. 2.2). How could people from so many countries still be interested in bows and arrows? (Perhaps eccentric people, and again in medieval costume—see fig. 6.2.) Why spend hard-earned money on mechanical clocks when a cheap digital watch tells the time much more accurately? And then there are those strange folk who, decked out in oily boiler suits, spend their free time assembling noisy, dirty, and pointless steam engines, which sometimes don't even move except for the funny whirly brass flyballs (fig. 6.3). How can a Frenchman earn a living making counterpoise

FIG. 6.3. *A 1912 steam traction engine, maintained by enthusiasts and exhibited at steam engine fairs. You can just see a flyball governor atop the engine. I am grateful to Rick Fairhurst for permission to reproduce this image.*

siege engines? Or a Hungarian making bows, or an Englishman making arrows? As we have seen, though, there are such people, and they point to a significant attraction for, and appreciation of, our machines in the minds of many people. The reason for this appreciation is, I suspect, partly cultural heritage (think of the Dutch and their windmills), and partly an admiration for clever devices and for those who made them.

Two of the five previous chapters have no biographical sketches. I might have said something about a modern-day bowyer, or millwright, or trebuchet maker, or about historical figures who chronicled the achievements of ancient bow or mill or trebuchet craftsmen. All these people have kept alive the memory of, and the many and varied skills required to build, these essential machines. But they did not contribute to the historical development. The names of the ancient geniuses who made the key technological developments are lost to us, because the bow and the siege engines originated in prehistory and ceased to be important before literacy became widespread. The waterwheel also is an ancient device, but it continued to be useful and so continued to be developed in relatively recent times. Consequently we know the names of a number of engineers and scientists whose aptitude and imagination led to improved waterwheel technology. Anchor escapements are a product of seventeenth-century Europe and their introduction and application is relatively well documented, in particular, in light of the "technological bottleneck" longitude problem. The centrifugal governor appeared at the end of the eighteenth century and also is well documented, again because of the large effort that went into solving a bottleneck problem. This time it was steam engine stability, and we know the names associated with the governor, and their contributions to it.

Our total ignorance of the earlier inventors, say the man or woman who invented the bow, compared with the wealth of detail we have concerning, for example, James Watt, makes it difficult for me to make comparisons. How clever and intelligent were all the people involved in the crucial technological developments outlined in this book? How much of a mental leap was required to form the key idea in each case? This question is not answered in any of the historical records, for any of our machines. By comparison we know exactly the mental leap made by James Watt when he first conceived of the separate condenser; we know because he has told us. But he did not invent the centrifugal governor, and I cannot find any reference to Thomas Mead's thought processes that led him to think of it. As for bows and arrows, well, we do not even know the millennium in which they were invented, let alone the name or circumstances of the inventor.

Another difficulty that arises concerns the environment in which the inventors lived. It is easier to invent something if you live in an age of innovation and inventions, surrounded by tools with which to build and test and refine your ideas. Others who will judge your innovations will have receptive minds if they share such an environment. But if you live in a society or an age in which the very *idea* of invention is unknown or unwelcome, then you have extra hurdles to jump. Sir Isaac lived at the beginning of the scientific revolution (indeed, it *was* the beginning because he was there), even so he had to invent much mathematics before he could give us gravity. Much more basic than that, though, he had to invent new words and concepts. The very idea of *speed,* so familiar to us, was not a concept that meant much to people in Newton's day (see Gleick, in Further Reading). The only unit of speed was the knot, used for ships at sea. So Sir Isaac had to set down foundations before he could build. Even so, he had something already to build on, though less than his successors. He said: "If I have seen further, it is by standing on the shoulders of giants." This may have been a jibe aimed at the diminutive Hooke, but it also referred to the work of others who preceded and influenced Newton. In turn, succeeding generations of scientists and engineers would stand on Newton's broad shoulders.[3]

So how can we compare the person who invented bows with the man who invented anchor escapements? We cannot, really, so I will not compare our inventors. An invention belongs to its own time. (What use would an anchor escapement be to a Neolithic hunter?) To make some judgment about how clever or innovative our five inventors have been, we must consider the times in which they lived. I will make a few observations and then, by and large, will leave it for you to decide how smart the inventors must have been.

The Inventors

YOU ARE UG, a stone-age cave dweller, probably in Africa. You have just returned home from hunting to your entry-level condo in your gated community. You haven't thought about complaining to the condo builder about the

---

3. Including myself, when writing the physics papers that form the technical backbone of this book.

lack of draft proofing, because nature built your cave, and lawyers haven't been invented yet. You sit by the latest state-of-the-art home heating device, known colloquially as a "fire." You haven't thought to worry about violating patent law on the use of this new device, because patents haven't been invented yet. What you *do* think about is how badly the hunt had proceeded and what slim pickings you brought home for your family: one snared rabbit, and one demented weasel who decided to attack you rather than run away. As you sit by the fireside, nursing your badly bitten nose, you cast an eye covetously at a herd of passing elk. Perhaps the idea of a bow and arrow comes to you as you consider the *atlatl* (spear thrower); the springy wood (and the extended lever arm) enables spears to be thrown further than they can be thrown by hand. Perhaps you watched Mrs. Ug light the fire with a friction drill. She used to rub this stick between her hands, but since she borrowed Mrs. Grunt's fire bow, she can light fires much more quickly. Maybe, once the fire is lit and you are enjoying barbecued weasel T-bone steaks,[4] you watch your kids playing with the fire bow and drill. Somehow, more or less by accident, the drill becomes an arrow. *Aha!* The drill flies only a few yards, but the idea is there.

In this fanciful scenario we can see a progression of ideas, of technology building upon itself. I think the bow and arrow probably came into Ug's head in something like the manner laid out here. Some level of abstraction was necessary, and several disparate ideas needed to come together: springy wood, fleet-footed prey, and projectile weapon.

Fast-forward 20 or 30 millennia. You are now Marcus Molinarius, an imaginary Roman citizen of, perhaps, the year 100 BCE. You are a millwright, and have constructed many undershot waterwheels that draw power from the sluggish streams and rivers that meander their way down the wide plane of Lombardy, let us say, in northern Italy. The principles of waterwheel operation were taught to you by your father, who was a miller descended from a long line of millers. Generations of empirical alterations and experimentation has led to an accumulated corpus of waterwheel lore that has increased and been passed down father to son. So now you find yourself ordered by the Roman army to build water-powered mills in an adjacent mountainous province, say Piedmont. No gentle streams here; you must construct a waterwheel on the side of a hill, harnessing the power of a rapid mountain stream. How

---

4. I guess that should be "t-bone," since weasels are not very large.

much of a mental leap is it to consider passing the water over the wheel, rather than under it? You have seen how bucket wheels draw water from a river, and felt the pull of the buckets as they were being emptied. Now the slope of the land makes overshot wheels easier than undershot wheels to construct. Your first design is simply a reverse bucket wheel, but on your second and third wheel you create improvements. This knowledge you pass down to your son, and so on down the centuries.

Now you have moved on seven centuries and moved east. I cannot provide you with a name, because I don't know if it should be Chinese, Central Asian, or Middle Eastern. You are in charge of a siege engine crew. You know all about traction trebuchets—those lever-arm engines that are powered by many men pulling ropes attached to the short lever arm. The fortress you are besieging is strong. Your task is made more difficult by staff problems: not enough crewmen to pull the ropes to launch the heavy missiles demanded by your commander. Or perhaps they are untrained, and unable to coordinate their efforts in the manner required to produce a long-range shot, or to hit the target consistently. You send your tired and thirsty crew to the river, to collect water. They raise the water with a *shaduf.* This is an ancient device that consists of a lever perched atop a fixed pole. The long lever arm has a bucket at one end, to collect the water, while to the short arm is fixed a balancing counterweight. You look at the shaduf raising water, and you look back at your traction trebuchet. How much of a mental leap is needed here? Your first counterpoise engine is a hybrid. A counterweight balances the projectile, while your crew still manipulates the ropes. This quickly leads to a reduced traction crew—your fittest and most coordinated men—and an increased counterweight. The rest of the crew simply raises the counterweight. It doesn't take long to realize that the traction crew can be disbanded, since the counterweight, once raised, can do all the work of launching the projectiles. The mangonel is born. Generations later, after much refinement of mangonel dimensions to obtain maximum range, one of your successors hits upon a hinged counterweight. Perhaps this started out by accident; perhaps a suspended bucket counterweight is cheaper, since it could be filled with rubble or sand. (A fixed counterweight would have to be formed from shaped stone or metal weights.) More refinements down the generations, and the trebuchet arises. It is a house built of many bricks by many builders, not hewn from a single marble block by one master craftsman.

Once again you move forward in time, to England in 1670, give or take a year. This time I cannot permit you to be the inventor, because we know the name of the man who gave us the anchor escapement. It was either Robert Hooke or William Clement. The idea of escapements had been around since the Middle Ages, in the form of the crown wheel.[5] The need for clock accuracy drove clockmakers to consider radically different regulator mechanisms and, from somewhere, the anchor escapement appeared. This new device cannot have been the result of many generations of knowledge passed down by word of mouth, except that the inventor needed to understand horology, in general, and the escapement principles, in particular. The anchor escapement arose in one person's mind as a specially invented solution to a specific problem. Because there was no obvious evolutionary development, and because I am not a clockmaker, it is difficult for me to reconstruct the mental processes of Clement/Hooke. So it seems to me that this was an *Aha!* invention; the product of a clever mind taking a large step forward all at once, rather than many minds advancing through many small steps.

Now you are in the year 1787, but still in England. Your friend Thomas Mead is about to invent the centrifugal governor. He may have invented it some years earlier, but 1787 was the year he was granted a patent. Bennett (see Further Reading) tells us that Watt had gathered evidence in preparation for patent litigation, suggesting that the centrifugal governor was used widely before Mead obtained his patent but, crucially for Mead, our friend Smeaton had not mentioned it in his windmill design of 1782. Whatever the truth of the matter, someone developed the centrifugal governor at about this time, and we may as well give Thomas Mead the benefit of the doubt, given his patent. So, from where did Mead get his idea? He was a mechanic and miller, so he knew his way around windmills. He will also have known about the idea of feedback control, since this principle had already been used in windmills, through the fantail direction controller, introduced earlier that century. But this does not tell us how Mead got hold of the flyball governor idea. Perhaps it was another *Aha!*—after all Mead lived in an age of innovation and mechanical awareness, and so he had the mindset for intentionally seeking improvement. This is not enough, though, to explain his innovation: he must have been an ingenious man.

---

5. And we have *no idea* who the inventive genius was, or how he came up with the crown and verge escapement.

And finally . . .

I HAVE NOW completed my elucidation of the bow and arrow, waterwheel and windmill, counterweight siege engine, anchor escapement, and centrifugal governor. I hope to have convinced you of their importance to human development, and of the genius, or at least inventive flair, of the known and unknown people associated with these interesting *ingenia*. There are plenty of other examples of cleverly designed machines. The common mechanical mousetrap is an ingenious device, but it does not merit a place here because it has not changed the world. (Of course this does not necessarily mean that the mousetrap inventor is less brilliant than Thomas Mead or Robert Hooke.)

Maybe *you* are brilliant. Perhaps you have a waterwheel, or at least a mousetrap, within you. I hope that you are Betty Jones, say, newly graduated from high school in Baltimore, and that this book has awakened in you an appreciation of physics or engineering, so you will go on to Johns Hopkins University and obtain a degree in control engineering. Or you are Dave Hernandez, a retired middle manager in Birmingham (Alabama or England or . . .) who now realizes that he is passionate about clock escapements, not stock indexing, and so designs a new escapement of coruscating, dazzling, shining originality. (Robert Hooke would probably want to sue you.) I would also be pleased if you turn out to be an auto mechanic who realizes after chapter 1 that you were really cut out to be a Parthian horse archer. Or after chapter 3 you knew that you were meant to be a trebuchet *ingeniator,* who has been trapped in the body of a software engineer all these years. Don't give up the day job, guys, but explore some more of the history and technology of these ingenious devices.

FURTHER READING

The reference lists at the end of each chapter cover both technical and historical aspects of the machines that form the subject matter of this book. For the general reader who would like to go into these areas in a little more depth, I recommend the following books, which are both accessible and authoritative. Most of them appear in more than one of the reference lists, such is their breadth. Here they are, listed alphabetically by author's name.

STUART BENNETT, *A History of Control Engineering 1800–1930*
An excellent introduction to the subject spawned by the centrifugal governor. With plenty of historical details, including diagrams of Thomas Mead's governor application and many other governor designs, this book sets the achievement of early control engineers into context, and shows how control engineering evolved internationally.

Bennett does not hold back on the mathematics, but these sections can be omitted by the timid.

ERIC BRUTON, *The History of Clocks and Watches*
A coffee table book, with many glossy and beautiful pictures of clocks and watches. A smaller paperback version is also available. This history of mechanical clock development includes nonmathematical explanations of the major innovations, with helpful diagrams. There is also a section on astronomical timekeeping.

JAMES GLEICK, *Isaac Newton*
A short but comprehensive and very incisive biography of the enigmatic, tormented Sir Isaac. Gleick knows his subject and weaves the whole story of Newton, not only his scientific accomplishments. This book is particularly good in conveying the (lack of) scientific base upon which Newton built his theories. Also included are details, newly uncovered and perhaps controversial, of Newton's extreme religious views; his copious writings on this subject far exceeded his scientific output.

JOHN G. LANDELS, *Engineering in the Ancient World*
Quirky and perhaps difficult to get a hold of, this book is a treasure for those of us who like ancient technology. The author is a classicist who understands engineering ideas, and he takes us through the ancient Greek and Roman technological world: power generation, pumps, cranes, catapults, and transport. The explanations are intuitive and nonmathematical, with plenty of helpful diagrams.

STEPHEN F. MASON, *A History of the Sciences*
A text that is over forty years old and out of print. However, it's much in demand and is available as a used book. This book has been described as "by far the best one-volume account in English of the development of the natural sciences," and I endorse that view. In 600 pages the author (a scientist) covers all of science and technology from the beginning to the mid-twentieth century. The coverage is necessarily terse, but all the major developments (such as the origins of Watt's steam engine) are here.

DAVA SOBEL, *Longitude*
A deservedly popular account of John Harrison and his struggle with the Longitude Board. Sobel emphasizes the history, rather than the technology. She is perhaps a little skimpy on the alternatives to longitude estimation, and on the contributions of non-English clockmakers, but her book is well worth reading for an appreciation of Harrison's genius and persistence against exasperating adversity.

JENNY UGLOW, *The Lunar Men*
A splendid and detailed account of the Lunar Society, this book describes the characters and times of these innovators (some well known and some not, but all influential in several fields) who shaped the dawn of the industrial revolution in England. There is a wealth of information on a dozen creative people, such as Josiah Wedgwood the potter, Matthew Boulton, and James Watt, and their friendships and rivalries during the last quarter of the eighteenth century.

ABBOTT PAYSON USHER, *A History of Mechanical Inventions*
Another old text, and still in print, this book provides a detailed but non-mathematical description of inventions from antiquity to the nineteenth century. The rather dense academic style is a small price to pay for the wealth of detail about waterwheels and windmills, clepsydra and mechanical clocks, printing and textile machinery, machine tools and the rise of the factory system. Usher also discusses the mental processes and social aspects involved in realizing novel ideas.